Advances in Logistics Engineering

Edited by Ágota Bányai

Published in London, United Kingdom

Advances in Logistics Engineering
http://dx.doi.org/10.5772/intechopen.1001754
Edited by Ágota Bányai

Contributors
Akunna Oledinma, Anjali Awasthi, Anthony E. Ezenwa, Ayoub Chakroun, Ágota Bányai Christopher Ikeogu, Dennis Sebata, Felix-Eduardo Bueno-Pascual, James H. Bookbinder, Martin Osei, Nidhal Rezg, Nze Ibeawuchi, Obiajulu Ede, Saravanan Natarajan, Xinyue Ren, Yonglin Ren

First published in London, United Kingdom, 2024 by IntechOpen
IntechOpen is the global imprint of INTECHOPEN LIMITED, registered in England and Wales, registration number: 11086078, 167-169 Great Portland Street, London, W1W 5PF, United Kingdom

British Library Cataloguing-in-Publication Data
A catalogue record for this book is available from the British Library

Additional hard and PDF copies can be obtained from orders@intechopen.com

Advances in Logistics Engineering
Edited by Ágota Bányai
p. cm.

This title is part of the Industrial Engineering and Management Book Series, Volume 8
Topic: Production Engineering
Series Editor: Fausto Pedro Garcia Marquez
Topic Editor: Orhan Korhan

Print ISBN 978-0-85466-320-0
Online ISBN 978-0-85466-319-4
eBook (PDF) ISBN 978-0-85466-321-7
ISSN 3029-0511

For EU product safety concerns:
IN TECH d.o.o., Prolaz Marije Krucifikse Kozulić 3, 51000 Rijeka, Croatia,
info@intechopen.com or visit our website at intechopen.com.

IntechOpen Book Series

Industrial Engineering and Management

Volume 8

Aims and Scope of the Series

Industrial Engineering and Management (IEM) is a discipline that focuses on optimizing complex processes and systems within various industries. It involves the integration of engineering, business, economics, mathematics, and behavioral sciences to improve efficiency, productivity, quality, and overall performance in organizations. Key aspects of Industrial Engineering and Management include: Process Optimization; System Analysis and Design; Quality Control and Management; Supply Chain Management; Operations Management; Human Factors and Ergonomics; Project Management; Cost Analysis and Financial Management; Decision Analysis.

Overall, Industrial Engineering and Management aims to optimize resources, improve processes, enhance productivity, and ensure the effective and efficient utilization of all elements involved in the production or delivery of goods and services. It is crucial in today's competitive business environment for organizations to stay efficient and competitive.

Production Engineering and Operational Excellence are fields of study and practices that focus on optimizing and improving the manufacturing and production processes within an organization. It combines principles from engineering, management, and operational strategies to enhance productivity, efficiency, quality, safety, and sustainability in the production of goods and services.

Here are the key components of Production Engineering and Operational Excellence: Process Optimization; Operational Excellence; Manufacturing Systems Design; Quality Management; Supply Chain Optimization; Production Planning and Scheduling; Automation and Technology Integration; Health, Safety, and Environmental Management; Cost Management; Performance Measurement and Key Performance Indicators (KPIs); Continuous Improvement and Innovation.

Production Engineering and Operational Excellence are crucial for organizations aiming to stay competitive in the global market by achieving high levels of efficiency, quality, and customer satisfaction while optimizing resources and minimizing waste. It is a multidisciplinary approach that encompasses engineering principles, management strategies, and the effective use of technology to drive operational success.

Meet the Series Editor

Fausto Pedro Garcia Marquez is a Full Professor at UCLM, Spain, with accreditation since 2013. He also holds the position of Honorary Senior Research Fellow at Birmingham University, UK, and serves as a Lecturer at the Postgraduate European Institute. In addition to these roles, Fausto has experience as a Senior Manager at Accenture from 2013 to 2014. He earned his European Ph.D. with the highest distinction. Throughout his career, Fausto has received numerous awards and honors. These include the Nominate Prize (2022), Gran Maestre (2022), Grand Prize (2021), Runner Prize (2020), and Advancement Prize (2018), as well as Runner (2015), Advancement (2013), and Silver (2012) by the International Society of Management Science and Engineering Management (ICMSEM). He was also the recipient of the First International Business Ideas Competition 2017 Award. Fausto's contributions extend to academic publishing, with over 242 papers to his name. Notably, his work has been recognized in journals like "Applied Energy" (Q1, IF 9.746, Best Paper 2020) and "Renewable Energy" (Q1, IF 8.001, Best Paper 2014). His affiliations include the editorial and authorship roles in more than 50 books, with publications through respected publishers such as Elsevier, Springer, Pearson, Mc-GrawHill, IntechOpen, IGI, Marcombo, and AlfaOmega. He has authored over 100 international chapters and holds 6 patents. Fausto serves as the Editor of 5 International Journals and is a Committee Member for more than 70 International Conferences. His research portfolio encompasses being the Principal Investigator in 4 European Projects, 8 National Projects, and participating in over 150 projects involving universities and companies. His areas of expertise and research interests span Artificial Intelligence, Maintenance, Management, Renewable Energy, Transport, Advanced Analytics, and Data Science. Fausto is a recognized Expert in the European Union in AI4People (EISMD) and ESF. He also serves as the Director of www.ingeniumgroup.eu, holds the status of Senior Member at IEEE since 2021, and has been honored as an Honorary Member of the Research Council of the Indian Institute of Finance since 2021. Fausto is also the Committee Chair of The International Society for Management Science and Engineering Management (ISMSEM) since 2020.

Meet the Volume Editor

Ágota Bányai was awarded an MSc degree in mechanical engineering from the University of Miskolc, Hungary, in 1993 and a Ph.D. degree in engineering sciences (Thesis title: "Design of Just in Time Systems") from the same institution in 1999. She is currently an associate professor in the Institute of Logistics at the University of Miskolc. She has more than 25 years of teaching and research experience in the design and control of logistic systems and supply chain management, with special emphasis on purchasing and distribution. She has worked in academia for nearly 25 years and has published over 200 research papers, book chapters, and conference proceedings.

Contents

Preface

This book offers a selection of chapters on supply chain solutions, promoting new research results in the field. The book covers seven topics, determined by the theoretical and practical aspects of logistics engineering.

Chapter 1, "Introductory Chapter: Advances in Logistics Engineering", focuses on the potentials in procurement, in-plant service and material handling, inventory management, distributions, facility location and layout design, traffic, transportation, and route planning. The chapter highlights that the Fourth Industrial Revolution has introduced innovative solutions and technologies that significantly improve the efficiency of global logistics and supply chain operations. These cutting-edge logistics technologies are driving processes that are cost-effective, sustainable, transparent, and flexible.

Chapter 2, "Enhanced Cross-Dock Productivity: Combining Self-Driving Vehicles with Forklifts", focuses on the replacement of some or many forklifts by self-driving vehicles. The impact of the proposed semiautomatic material-handling system on productivity is analyzed. Financial aspects of the proposed solution are also discussed. The developed simulation models of different cross-docking solutions are analyzed, and a near-optimal set of material-handling equipment is also defined.

Chapter 3, "Forces Transforming Transport and Logistics into Smarter Sustainable Systems", extensively discusses the impact of digitalization, shifts in international trade, enterprise-software-driven process changes, changes in markets' domestic commerce, and machine-driven process changes on the transformation of modern transport and logistics. The approach focuses on the improvement of operations, increasing performance, efficiency, and effectiveness, but sustainability aspects are also taken into consideration.

Chapter 4, "Advanced Sustainable Logistics with HSR for the Development in Great Montreal Area", addresses the impact of high-speed railway on the regional development in terms of trade centered on the Great Montreal area. Mathematical models based on the gravity model are extensively discussed. Based on the scenario analyses, the research discusses the impact on the economic development resulting from HSR.

Chapter 5, "Application of Machine Learning for Predictive and Prognostic Reliability in Flexible Shop Floor", focuses on the use of machine learning to predict the reliability of flexible workshops, while optimization of production management is also discussed. The chapter presents cost optimization models to prevent unplanned breakdowns, extend equipment lifespan, optimize spare parts usage, and ensure smooth, uninterrupted production in the flexible workshop.

Chapter 6, "Navigating Sustainability: A Conceptual Exploration of Road Freight Decarbonisation", contributes to highlighting the importance of embracing alternative fuels, specifically natural gas and hydrogen, to decarbonize road freight operations. The chapter urges interdisciplinary research and collaboration to address regional and industry-specific challenges in road freight, emphasizing technological innovation, policy frameworks, and economic considerations for sustainable global practices.

Chapter 7, "Investigating How Logistics Operations Affect an Organization's Financial Performance", emphasizes the critical role of efficient logistics in a company's financial success, highlighting the need for improved financial control over product and service transport. It provides guidance on streamlining logistics processes to enhance profitability through timely deliveries, reduced inventory costs, and savings on transportation expenses. By leveraging technologies like AI and machine learning, businesses can use data analytics to optimize their supply chains and strengthen supplier relationships for better overall performance.

The aim of this book is to help students as well as managers and researchers to understand and appreciate the advances in logistics engineering. The editors thank the chapter authors for their scientific contributions. The chapters were edited and published following a rigorous selection process. I also wish to thank and acknowledge the many individuals who helped us throughout the editorial process that made this book possible.

Ágota Bányai
Institute of Logistics,
University of Miskolc,
Miskolc, Hungary

Chapter 1

Introductory Chapter: Advances in Logistics Engineering

Ágota Bányai

1. Introduction

Today, logistics processes are becoming increasingly important in economic processes. One of the main reasons for this is that logistics processes have a significant impact on key performance indicators (KPIs) at each stage of the value chain. These indicate that the competences of logistics processes are basically focused on the areas of procurement, production, distribution, and recycling and that these four sub-processes can be considered as the horizontal integration of logistics processes. These four sub-processes constitute the circular economy, and it follows that the proper design of logistics processes not only helps to increase cost efficiency but can also contribute greatly to the development of sustainable, environmentally friendly value chains. In addition to the horizontal integration of logistics processes, we can also talk about vertical integration, where logistics processes can be defined as the integration of strategic, tactical, and operational levels (**Figure 1**).

Logistics engineering is a special field of logistics that can be understood at all three levels. It includes the material handling machines and material handling equipment, IT and telecommunication solutions, planning, and operational algorithms needed to service production and service activities. Within the frame of this introductory chapter, the most important research fields of advanced logistics engineering are discussed, focusing on the potential solutions in the Industry 4.0 era.

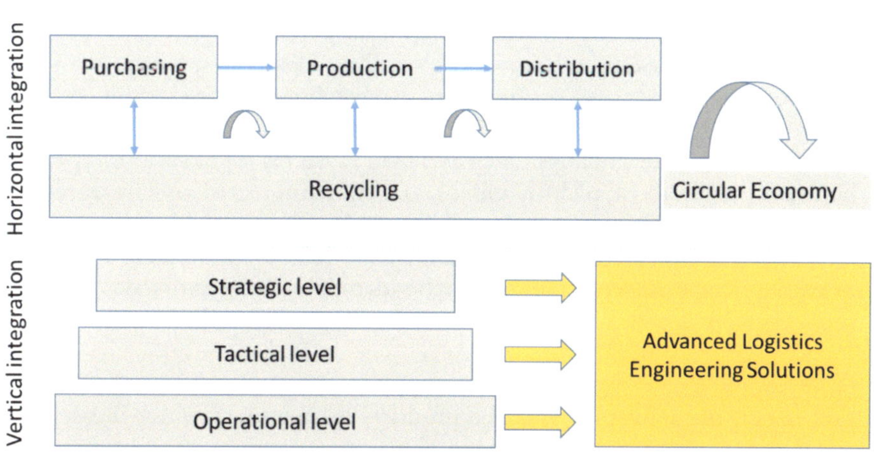

Figure 1.
Horizontal and vertical integration of logistics.

2. Advances in logistics engineering

Logistics engineering focuses on many areas of manufacturing and service companies, where increasing efficiency is a prerequisite for fulfilling customers' demands at the highest possible level:

- Procurement: The procurement of the tools, components, and raw materials necessary for the proper operation of production and service processes is a fundamental prerequisite for the operation of the value chain. Procurement is a key element in the operation of the value chain. Procurement is responsible for providing resources, components, and raw materials for production and service processes to ensure that customer needs can be met in a cost-effective and sustainable manner. A new trend in procurement is the use of automation and artificial intelligence, which plays an important role in supporting procurement tasks related to uncertain customer needs. Another important trend is blockchain, which can ensure transparency of processes and increase trust. Chain of custody [1] and batch tracking [2] are also important application fields of blockchain technology in purchasing.

- In-plant service, materials handling: For production and service activities, it is essential to perform material handling tasks to ensure that materials are moved to the right place and handled to meet customer needs. Such material handling tasks are loading, unloading, transportation, packaging, and storage. A new trend in the field of in-plant transport is the emergence of autonomous transport systems that are essentially AGV-based [3] and can automatically adapt to dynamically changing in-plant material handling tasks. A good example is the AGV-based in-plant material handling system in matrix manufacturing [4], where a large number of AGVs in an AGV pool autonomously determine the optimal material supply sequence through real-time optimization, which is particularly challenging because in matrix manufacturing the production and logistics processes are decoupled [5].

- Warehousing, storage, and inventory: Inventory-related processes are increasingly the focus of attention, as stocks do not essentially create surplus value, so it makes sense to reduce stocks and the cost of stockpiling. However, this can have a major impact on the service level, as the ability to meet dynamically changing customer needs can be greatly increased through inventories. In the area of inventory management, modern technologies such as picking systems [6] of various types (pick-by-voice, pick-by-light, pick-by-video), and fully automated satellite warehouse solutions [7, 8] can help to increase efficiency. An interesting area of warehousing is the use of technical solutions that can replace the static warehouse structure with a flexible warehouse structure that can be adapted to the quantity and size of the products to be stored, and to the forecasted storage needs [9].

- Distribution: One of the most important phases of production and service processes is the delivery of finished products to customers, which is essentially an operation with a high environmental impact. Therefore, in addition to cost efficiency, increasing sustainability is becoming increasingly important, and one important solution could be the development of integrated supply chains that aim to integrate first-mile and last-mile activities [10–12]. In distribution, a new energy-efficient solution could be the integration of traditional delivery and drone delivery activities [13], as well as the increasing use of parcel pick-up points [14], whereby the last part of the last mile is delivered to the customer.

- Facility location and layout design: An important factor influencing the efficiency of logistics processes is the development of an optimal layout, which reduces the necessary material handling distances and increases the availability of suppliers and customers. Layout tasks are basically NP-hard optimization problems [15] that can be solved by heuristic and metaheuristic methods. Thus, efficient, accurate, and fast optimization methods are becoming increasingly important in the design of layouts that define the structure of global supply chains and in-plant layouts. A new trend in logistics process management is the development of flexible layouts, one example being the matrix manufacturing system, where the layout is not flexible, but the flexible production units allow the layout to be flexible, as production tasks can be flexibly assigned to the technological resources in the matrix grid according to customer demand [16].

- Traffic, transportation, and route planning: Transportation is the part of logistics processes with the highest environmental impact. To reduce the environmental impact, it is necessary to design routes that minimize emissions. In addition to routing, another important means of reducing the emissions from transport is, of course, the choice of the right mode of transport, which is essentially shifting towards rail transport [17] and the use of e-trucks [18]. In passenger transport, there is a clear trend towards prioritizing public transport.

In developing logistics processes, it is essential to exploit the potential of Industry 4.0 technologies, so the use of advanced technologies and methods such as digital twin, big data solution, virtual and augmented reality, intelligent sensors and sensor networks, cloud computing, fog computing, edge computing, artificial intelligence, state-of-the-art identification technologies, autonomous machines, and robots, cobots.

The fourth industrial revolution is very different from the previous industrial revolutions, as the new ideas of Industry 4.0 do not require any new inventions. The Internet, artificial intelligence, cloud services, and similar technological breakthroughs already existed at the time of the third industrial revolution. Rather, Industry 4.0 is a new concept and vision, the main ambitions of which are

- Vertical integration, that is linking logistics, manufacturing, marketing, and all other activities and resources related to the product through various smart solutions, to create a customer-specific, pull manufacturing plant with a free flow of information.

- Horizontal integration, which brings business partners and customers closer together not only within countries but also on a global scale.

- Flexible factories that combine the advantages of customized manufacturing and mass production. This production method is called custom mass production.

These efforts can be achieved by connecting the equipment and devices in the factory into a single intelligent network. The main elements of Industry 4.0 integration are logistics, manufacturing, industrial design, quality management, product management, energy management, data collection, and data analytics.

Within the frame of the next part of this chapter, some typical research directions are highlighted, to show the focuses of advanced logistics engineering.

- Industry 4.0 technologies and the integration of Internet of Things solutions lead to improved production and service systems and processes, where the application of blockchain technologies can significantly increase the trust in the members of the supply chain, while transparency is also taken into consideration [19].

- The Industry 4.0 technologies have a great impact on automation, and the advanced technologies of materials handling and warehousing are also used in the field of development and improvement of warehousing and storage operations, as a research in the field of automated storage systems with shuttle shows [20].

- The design and operation of advanced logistics systems are generally based on simulation studies [21]. The design process of advanced logistics systems has three levels of design: the conceptional design of processes, the optimization of processes, and the simulation-based planning. In the third phase, 3D modeling technologies can be used to improve the visualization of logistics processes.

- Digitalization led to the transformation of conventional manufacturing and production processes into digital factories [22].

- The big data problems in production, services, and related global supply chain resulted complex decision-making problems, where machine learning and artificial intelligence technologies represent potential solutions [23].

- Empirical approaches play an important role in the design and operation of complex logistics solutions. An interesting approach is shown in [24] focusing on the application of entropy method and Analytic Hierarchy Process.

- Not only technology and system engineering, but also business is a significant influencing factor in logistics engineering. Economic and financial aspects become more and more important in logistics engineering [25], while sustainability is also a core problem.

- Real-time decision-making can be defined as a core competence in advanced logistics solutions, where digital twin [26] and improved micro-sensors, intelligent sensors, sensor networks, and edge computing can support real-time decision-making to fulfill dynamically changing demands of customers.

- The design of advanced logistics processes includes a wide range of design aspects, including routing [27], facility location, routing, scheduling, assignment, or clustering. In the case of complex logistics solutions, integrated optimization models are used and their solutions are generally based on metaheuristic and artificial intelligence.

- The transportation sector is also significantly influenced by globalization and economic growth, and the economic growth and the increased complexity of supply chain solutions can controlled by appropriate Industry 4.0 technologies [28]. This influence can be recognized both in the domestic, industrial, and transportation sector [29, 30].

- The COVID-19 pandemic has led to new approaches to design and operation of logistics systems, especially in the field of supply chain. Digitalization and e-solutions become more and more significant [31].

3. Conclusions

Advanced technical and systems engineering solutions are playing an increasingly important role in the logistics processes associated with the operation of production and service systems. The Fourth Industrial Revolution has brought new solutions and technologies to enhance the efficiency of increasingly global logistics processes and supply chains. These advanced logistics technologies are leading to cost-effective, sustainable, transparent, and flexible processes.

Conflict of interest

The author declare no conflict of interest.

Author details

Ágota Bányai
University of Miskolc, Institute of Logistics, Miskolc, Hungary

*Address all correspondence to: agota.banyaine@uni-miskolc.hu

References

[1] Patil H, Kohli RK, Puri S, Puri P. Potential applicability of blockchain technology in the maintenance of chain of custody in forensic casework. Egyptian Journal of Forensic Sciences. 2024;**14**(1):12. DOI: 10.1186/s41935-023-00383-w

[2] Panda SK, Satapathy SC. Drug traceability and transparency in medical supply chain using blockchain for easing the process and creating trust between stakeholders and consumers. Personal and Ubiquitous Computing. 2024;**28**(1):75-91. DOI: 10.1007/s00779-021-01588-3

[3] Chen X, Liu S, Zhao J, Wu H, Xian J, Montewka J. Autonomous port management based AGV path planning and optimization via an ensemble reinforcement learning framework. Ocean and Coastal Management. 2024;**251**:107087. DOI: 10.1016/j.ocecoaman.2024.107087

[4] Mohr J, Schmidtke N, Behrendt F. Design of Matrix Production Systems: New demands on factory planning methods. Procedia Computer Science. 2024;**232**:1972-1981. DOI: 10.1016/j.procs.2024.02.019

[5] Ranke D, Bauernhansl T. Modeling of a matrix production system for simulation to predict material demand. IFIP Advances in Information and Communication Technology. 2023;**691**:676-690. DOI: 10.1007/978-3-031-43670-3_47

[6] De Vries J, De Koster R, Stam D. Exploring the role of picker personality in predicting picking performance with pick by voice, pick to light and RF-terminal picking. International Journal of Production Research. 2016;**54**(8):2260-2274. DOI: 10.1080/00207543.2015.1064184

[7] Shu B, Pei F, Zheng K, Yu M. LIRP optimization of cold chain logistics in satellite warehouse mode of supermarket chains. Journal of Intelligent and Fuzzy Systems. 2021;**41**(4):4825-4839. DOI: 10.3233/JIFS-189968

[8] Reis ADC, Oliveira RP, Ferreira LAF, de Souza CG. Lessons from a surgical center satellite warehouse in a large brazilian public hospital. Healthcare (Switzerland). 2021;**9**(3):297. DOI: 10.3390/healthcare9030297

[9] He M, Guan Z, Hou G, Wang X. A novel parts-to-picker system with buffer racks and access racks in flexible warehousing systems. Sustainability (Switzerland). 2024;**16**(4):1388. DOI: 10.3390/su16041388

[10] Bányai T. Real-time decision making in first mile and last mile logistics: How smart scheduling affects energy efficiency of hyperconnected supply chain solutions. Energies. 2018;**11**(7):1833. DOI: 10.3390/en11071833

[11] Bányai T. Impact of the integration of first-mile and last-mile drone-based operations from trucks on energy efficiency and the environment. Drones. 2022;**6**(9):249. DOI: 10.3390/drones6090249

[12] Bányai T, Illés B, Bányai Á. Smart scheduling: An integrated first mile and last mile supply approach. Complexity. 2018;**2018**:5180156. DOI: 10.1155/2018/5180156

[13] Dukkanci O, Campbell JF, Kara BY. Facility location decisions for drone

delivery: A literature review. European Journal of Operational Research. 2024;**316**(2):397-418. DOI: 10.1016/j.ejor.2023.10.036

[14] Niemeijer R, Buijs P. A greener last mile: Analyzing the carbon emission impact of pickup points in last-mile parcel delivery. Renewable and Sustainable Energy Reviews. 2023;**186**:113630. DOI: 10.1016/j.rser.2023.113630

[15] Paschos VT. An overview on polynomial approximation of NP-hard problems. Yugoslav Journal of Operations Research. 2009;**19**(1):3-40. DOI: 10.2298/yjor0901003p

[16] Bányai T. Optimization of material supply in smart manufacturing environment: A metaheuristic approach for matrix production. Machines. 2021;**9**(10):220. DOI: 10.3390/machines9100220

[17] Danesi A, Ongari D, Poliziani C, Rupi F. Evolution of the road and rail transport of goods in european countries before and after the financial crises. Communications - Scientific Letters of the University of Žilina. 2019;**21**(4):3-12

[18] Liimatainen H, van Vliet O, Aplyn D. The potential of electric trucks – An international commodity-level analysis. Applied Energy. 2019;**236**:804-814. DOI: 10.1016/j.apenergy.2018.12.017

[19] Shen A. Design of internet of things service system for logistics engineering by using the blockchain technology. International Journal of Grid and Utility Computing. 2023;**14**(2-3):182-190. DOI: 10.1504/IJGUC.2023.131003

[20] Lerher T, Bencak P. Advanced Technologies in Logistics Engineering: Automated storage systems with shuttles integrated with hoisted carriage. Tehnicki Glasnik. 2022;**16**(3):336-342. DOI: 10.31803/tg-20220509104609

[21] Liu D, Pan Y, Li L. Logistics engineering simulation using computer 3D Modeling technology. Journal of Physics: Conference Series. 2022;**2143**:1. DOI: 10.1088/1742-6596/2143/1/012018

[22] Michlowicz E. Logistics engineering and industry 4.0 and digital factory. Archives of Transport. 2021;**57**(1):59-72. DOI: 10.5604/01.3001.0014.7484

[23] Bu S. Logistics engineering optimization based on machine learning and artificial intelligence technology. Journal of Intelligent and Fuzzy Systems. 2021;**40**(2):2505-2516. DOI: 10.3233/JIFS-189244

[24] Zhang YR, Cao WF. Empirical research on logistics engineering based on AHP and entropy method. Applied Mechanics and Materials. 2014;**668-669**:1663-1666. DOI: 10.4028/www.scientific.net/AMM.668-669.1663

[25] Klaus P. Business logistics and logistics engineering: The rocky road to a unified science of logistics. Logistics Research. 2010;**2**(1):1. DOI: 10.1007/s12159-010-0027-4

[26] Martínez-Gutiérrez A, Díez-González J, Ferrero-Guillén R, Verde P, Álvarez R. Digital twin for automatic transportation in industry 4.0. Sensors. 2021;**21**:10. DOI: 10.3390/s21103344

[27] Twarog S, Szwarc K, Wronka-Pospiech M, Dobrowolska M, Urbanek A. Multiple probabilistic traveling salesman problem in the coordination of drug transportation-In the context of sustainability goals and industry 4.0. PLoS One. 2021;**16**:3. DOI: 10.1371/journal.pone.0249077

[28] Adedoyin FF, Bekun FV, Driha O, Balsalobre-Lorente D. The effects of air transportation, energy, ICT and FDI on economic growth in the industry 4.0 era: Evidence from the United States. Technological Forecasting and Social Change. 2020;**160**:120297. DOI: 10.1016/j.techfore.2020.120297

[29] Laxmi DB, Bhaumik M, Jayan N, Sukanya B. A study on influence of industry 4.0 in the domestic, industrial and transportation sector. Journal of Critical Reviews. 2020;**7**(7):809-811. DOI: 10.31838/jcr.07.07.148

[30] Letnik T, Hanžič K, Luppino G, Mencinger M. Impact of logistics trends on freight transport development in urban areas. Sustainability. 2022;**14**:24. DOI: 10.3390/su142416551

[31] Zondervan NA, Tolentino-Zondervan F, Moeke D. Logistics trends and innovations in response to COVID-19 pandemic: An analysis using text mining. PRO. 2022;**10**:12. DOI: 10.3390/pr10122667

Chapter 2

Enhanced Cross-Dock Productivity: Combining Self-Driving Vehicles with Forklifts

Saravanan Natarajan and James H. Bookbinder

Abstract

A cross-dock (CD) in a supply chain avoids storing goods that would be picked for orders soon after. Vehicles inbound to the CD are unloaded and their contents are re-sorted. Appropriate items are then loaded within a short time on outbound vehicles for shipment to customers. The CD material handling operations of unloading, sorting and loading are typically done "manually", by forklifts with human operators. In this chapter, we consider the replacement of some or many forklifts by "Self-Driving Vehicles" (SDV). Can the resulting semi-automated material handling system attain the same or greater productivity as the fully manual system? At what cost (per unit of output)? We develop simulation models of two CDs, one purely manual and the other containing a mixture of forklifts and SDVs. Several CD performance measures are defined and estimated *via* simulation. For each CD, response surface methodology is employed to determine a near-optimal set of material handling equipment, when that CD is operated at a specified performance level.

Keywords: logistics, material handling, simulation and optimization, manual plus automation, cross-dock

1. Introduction

The increased variety of items demanded poses greater challenges in shipping the right quantities of those goods from the supplier to each customer. Often the demand for a particular product does not meet a full truckload (FTL). Suppliers use cross-dock facilities to improve outbound vehicle utilization.

Cross-docks coordinate shipments between particular suppliers and customers. Goods demanded by several retail centres or customers are sent from various suppliers to a central hub, the CD. Products received are then dispatched to customers within 24–48 hours of arrival at the cross-dock, having been sorted and consolidated using labour and MHE. Requirements of individual retail centres or customers are thus shipped in an almost Just-In-Time manner, without holding stocks at the CDs.

1.1 Material handling in a cross-dock facility

CD performance metrics include throughput rate (pallets processed per unit time), average truck turnaround time, MHE utilization, doorway or dock utilization, and shipping accuracy. Those are all directly influenced by the efficiency of material handling activities. MHE, for example hand pallet trucks and forklifts, are widely used in cross-dock facilities to process inbound items. However, recent advances in robotics for material handling, for example Automated Guided Vehicles (AGV) and Autonomous Mobile Robots or Self-Driving Vehicles (SDV), motivated our research on the scope of SDV in a CD.

Automated Guided Vehicles are programmed to move only on a dedicated path, making them well-suited for handling and material movement in a warehouse or other facility. The designated path is marked by paint or wire. SDVs, however, can move around an aisle or route by sensing any objects in their way. The ability of the SDV's central control system to choose optimal routes, and the SDV's obstacle-manoeuvring capabilities, makes Self-Driving Vehicles a potential alternative to forklifts. SDVs can thus be used in a CD to process pallets or goods received as unit loads.

A generic cross-dock facility, if operating manually, would require forklifts to sort and transfer those pallets between inbound and outbound trucks. A "semi-automated CD" [1] would process pallets with a mixture of forklifts and SDVs. Use of SDVs for cross-docking could perhaps reduce the variable operating cost, since less labour would likely be needed for such a semi-automated facility, compared to a manually operated CD. The present research financially compares a cross-dock operating with only forklifts to one operating with a mixture of forklifts and SDVs.

1.2 Solution methodology

To accommodate randomness in a CD material handling operation, we employed the discrete event simulation software ARENA. Two models were developed: *Simulation Model 1)* Forklift-only cross-dock (FL-only CD); and *Simulation Model 2)* Forklift-and-SDV cross-dock (FL-and-SDV CD). The models were built to be flexible and scalable. (Modelling assumptions and operating conditions are discussed in Section 3.) Independent simulation experiments were run with each model by varying the MHE quantities in offloading, sorting and loading. Response Surface Methodology enabled CD output performance metrics to be modeled in terms of MHE allocated to each process.

The latter prediction equations then permitted the formulation of an optimization problem. Its objective function is to minimize the operating cost of a CD facility, subject to constraints on the CD performance metrics we employ (throughput rate,

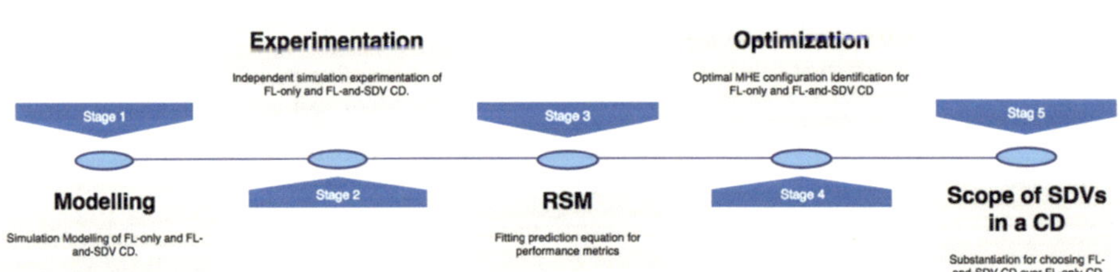

Figure 1.
Solution methodology.

throughput rate/MHE and MHE utilization rate). That optimization model was solved for desired throughput rates; the corresponding optimal MHE configurations were thus identified. Our overall solution framework is given in **Figure 1**.

2. Literature review

There are ample publications that treat an overall cross-docking operation. But only a few papers were found that focused on modelling and analysis of *floor-level* activities, with different mixtures of MHE. Fewer still optimized those material handling operations, directly or indirectly. In that light, we discuss the relevant papers available on the modelling of material handling in a CD, first with deterministic and then stochastic models. Unless stated otherwise, the MHE are forklifts.

2.1 Deterministic modelling of CD material handling events

Bartholdi and Gue [2] optimized the labour cost of an LTL cross-docking operation, subject to floor-space congestion, that is, the ratio between average pallet flow and the total area available in front of an outbound door. Their objective function minimized the expected costs of moving pallets in rectilinear distance plus waiting due to interference between MHE.

Gelareh et al. [3] developed new formulations for CD door assignment. While all models are deterministic, the transportation times and distances between inbound and outbound doors were carefully calculated. Wolff et al. [4], noted, as we have, that efficient operation of a cross-dock is reflected by how internal resources (labour, MHE) are employed. *Via* a (deterministic) mixed-integer programming model, those authors sought a feasible truck schedule carried out with minimal resources.

The preceding analyses of cross-dock material handling yielded optimal or near-optimal solutions of a deterministic model. However, optimality in *real* CD facilities must deal with high randomness in material handling operations [5]. Simulation modelling is next discussed.

2.2 Stochastic simulation of CD material handling

Magableh et al. [6] modeled a generic cross dock in which arriving trucks are assigned available doors based on FIFO. Using MHE, goods are then unloaded, sorted and loaded onto outbound vehicles for dispatch. Although this model represented the randomness in real-world CD material handling activities, the facility parameters (e.g. size, number of workers, MHE composition) were not varied or even specified explicitly. Liu and Takakuwa [7] minimized the labour cost of a retail distribution CD. Their simulation model treated stochastic operator movements, but randomness involved in actual MHE activities was ignored. Yang et al. [8] apply simulation to study the impacts on CD operation of the numbers of forklifts, door layout, and the freight mix for the case of purely manual MHE. Those authors analyzed the mean throughput rate/forklift and mean pallet-handling time in a CD. Forklift movements were modelled as they would occur in a real-world cross-dock, but optimal MHE levels were not identified.

Adewunmi and Aickelin [9] proposed a simulation and optimization framework covering CD order consolidation performed by 5–7 operators using –5 pieces of MHE.

Their Total Cost objective was evaluated by simulation within a relatively small solution space.

Torbali and Alpan [10] systematically reviewed models of cross-dock scheduling that take a real-time perspective.

Let us now turn to *simulation-optimization* problems. These are generally more complicated to solve than deterministic optimization models, due to the difficulty in objective function evaluation. The technique involves independent simulations of each feasible set within the optimization search space.

Rather than simulating all those feasible sets, researchers have built and analysed the *meta-models*[1] of a simulation model, to estimate the optimal solution of the simulation-optimization problem. Regression Analysis and Neural Networks have been widely used for such purposes [11].

Aickelin and Adewunmi [12] solved the cross-dock door assignment problem to reduce the distance traveled by MHE with pallets. Bottlenecks and delays involved in a real-world CD were modeled, but MHE performance was not studied, hence not optimized. Shi et al. [13] proposed a simulation- optimization framework to design a robust JIT-based CD. Suh [14] employed agent-based techniques and then simulation to model the overall product movement from suppliers to distributors *via* CD. The simulation model addressed uncertainties in waiting times and logistics inventories but did not include material handling activities.

2.3 Autonomous material handling

Only a few papers on applications of SDV for material handling purposes have been found, but we have identified research on AGV use in the factory, warehouse or CD-like facilities. Peixoto et al. [15] applied simulation modelling techniques to study the impact of a new automated MHS for warehouse order picking, with results compared to the performance of a manually operated facility. He and Prabhu [16] study a CD where AGVs enable outbound vehicles to get the proper packages directly from inbound trucks. They develop a queueing model to establish the CD's performance measures and determine the number of AGVs. Simulation is then used to check the model's accuracy.

Rupp et al. [17] treat the Hub-Arrival-Departure problem. A cross-dock is an important example, whereby each vehicle must be assigned two-time intervals: first for arrival and then for departure. A consolidation time is needed at the hub to sort, prepare and rearrange products for outbound dispatch. Because they do not deal with a CD as such, Rupp et al. do not drill down to the details of material handling.

Guo et al. [18] also emphasize the treatment at hubs via *parallel-machine scheduling*. In the case of material handling *via* forklifts, the goal is little or no waiting in the transfer of pallets from inbound to outbound vehicles. The two trucks must be given time windows with maximal overlap, in order that the forklift be able to smoothly transfer the goods. Again, it is the *details* of the MHE utilized that separate our research from others. We concentrate on how the pallets are actually moved and transferred (solely by forklift or by a mixture of forklift-and-SDV?). Articles such as [17] or [18] could furnish important inputs into deciding the door assignment of truck

---

[1] A meta-model (also called surrogate model) is a model of an already existing model. Meta-modelling is the process of generating such meta-models.

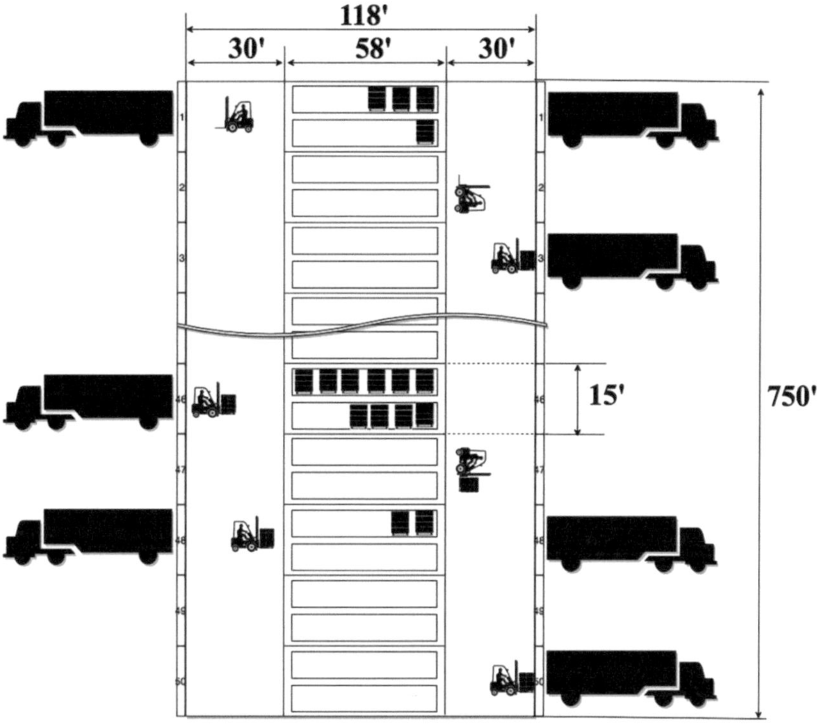

Figure 2.
Facility layout: forklift-only cross-dock facility.

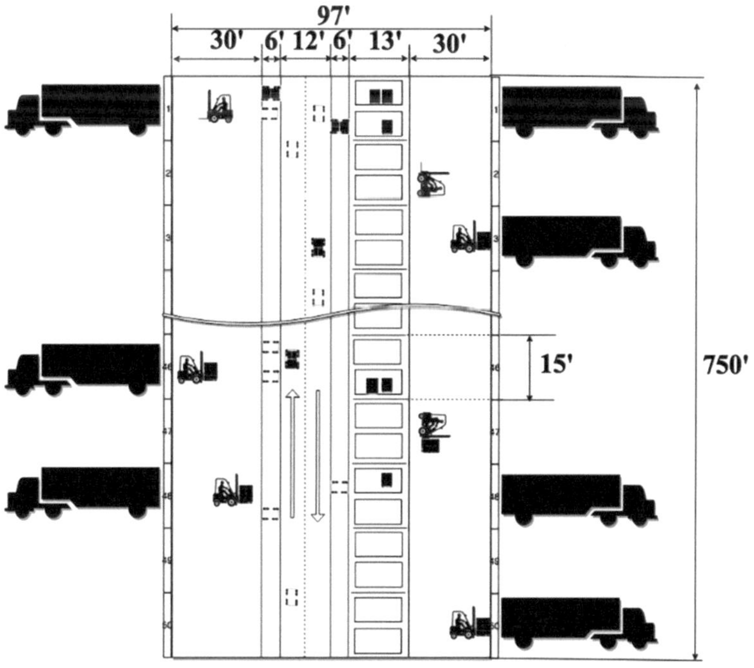

Figure 3.
Facility layout: forklift-and-self-driving vehicle cross-dock facility.

arrivals at the CD. However, our issues concern the particular MHE operations on the facility floor.

Sayed et al. [19] do consider the times required for within-CD handling, that is, for unloading, transfer and loading of products. Those authors stress the importance of recognizing a *spatial* dependence in estimation of the handling times, in terms of the "route" taken within the CD. Indeed, in our research, the cross-dock layout (**Figure 2** or **Figure 3**) implies such time and distance functionality. For us, the precise mixture of MHE will then determine the makespan, for a given truck arrival and departure pattern, and mixtures of goods involved.

Fragapane et al. [20] distinguish between centralization and decentralization. What we term "SDV", Self-Directed Vehicles, they call "AMR", or Autonomous Mobile Robots. The larger the number of AMRs in the system, the more likely that decision-making will be decentralized. Our system, with its adjustable mixture of forklifts and self-directed vehicles, employs centralized decision-making. Our simulation models thus enable smooth coordination between the MHE (SDV and forklifts). Our SDV (see Section 3.2 and **Table 1**) can do all that a forklift can: fetch, sort and transport. The real benefit of an SDV, of course, is that it requires no human operator. With these attributes of our automated MHE, it is reasonable for us to compare the capabilities of CDs with varying mixtures of forklifts and SDV.

We emphasize that SDVs permit stable manoeuvring around obstacles and smart decision-making to choose an optimal path. This makes their movement equivalent to

Parameter	Value
Model naming CD working/simulation run time	(IB-FL). (SDV). (OB-FL) 7.5 *hr/day*
Facility dimension	
Length Width	750 *ft.* 97 *ft.*
Trucks	
IB or OB truck inter-arrival time Pallets per IB or OB truck	0 *UNIF* (20, 26)
Docks	
Number of inbound doorways or docks	50
Number of outbound doorways or docks	50
OB Staging capacity	13 Pallets in 15 × 13 ft^2
Number of pallet holders	1
MHE speed	
Forklifts with pallet Forklifts without pallet SDV with or without pallets	316 *ft/min* or 1.61 *m/s* 548 *ft/min* or 2.78 *m/s* 316 *ft/min* or 1.61 *m/s*
Time delays	
IB or OB truck changeover time MHE travel time MHE time to manoeuver and pick up pallets MHE time to manoeuver and drop-off pallets	$\frac{TRIA(2,3,5)\ mins.}{manhattan\ Distance/Speed}$ *UNIF* (8, 12)*secs* *UNIF* (8, 12)*sec* attempts

Table 1.
Design parameters: forklift-and-SDV cross-dock facility.

the movement of manual forklifts in a free path. Therefore, simulation modelling of SDV movements requires techniques similar to those for simulating manual forklifts, not AGVs.

3. Simulation modelling

3.1 Cross-dock design

The volume of goods which can be processed in a CD is determined by its cross-dock geometry. How many inbound and outbound doors does it have? How large are the staging area, MHE aisle, MHE parking area, MHE charging stations and the measuring stations? We will focus on the CD's dedicated material handling region. That is, the number of doors and door size, facility shape, staging area and MHE aisle.

There are two types of doorways in a cross-dock. Goods from arriving vehicles are unloaded at the ***inbound doors***. Trucks will leave the CD at ***outbound doors***; their consolidated loads of products (received from inbound vehicles) are sorted by destination. For this research, a cross-dock is assumed to have a total of 100 doors, each 15 ft wide. The various CD shapes include I (long narrow rectangular), **L, H, U, T, H, X** and **E**. An **I**-shaped narrow rectangular cross-dock with 100 doors is assumed, based on the computational experiments on CD shapes of Bartholdi and Gue [21]. That is, 50 doors each on opposite sides of a facility, for a total length of 750 ft.

In an ideal cross-dock, materials received can be directly loaded onto outbound trucks from those inbound using MHE, when OB trucks are available, that is, parked at outbound doors. In cases where the destination vehicles are unavailable or goods require sorting and consolidation, the products are put in temporary storage, called a **Staging Area**, for further action. A CD facility with inadequate staging area suffers serious congestion and time delays; an excessive staging area would result in un-utilized floor space. The following are our assumptions on staging areas and aisles.

Forklift-only cross-dock facility (**Figure 2**):

Staging Area: Floor space of $58 \times 15 ft^2$ is dedicated for each pair of doors from opposite sides.

Forklift Aisle: Floor space of $30 \times 15 ft^2$ is dedicated in front of each door, for the forklifts to manoeuver, move between doors, sort goods and for truck loading or unloading

Forklift-and-self-driving vehicle cross-dock facility (**Figure 3**):

Staging Area: Floor space of $13 \times 15 ft^2$, allocated to each pair of doors on opposite sides. *Forklifts Aisle:* Floor space of $30 \times 15 ft^2$ is dedicated in front of each door for the forklifts to manoeuver, unload and load the trucks.

Self-Driving Vehicle Aisle: Floor space of $12 \times 150 ft^2$ is dedicated for the SDV to move back and forth across the CD. For each door, a floor space of $6 \times 15 ft^2$ is dedicated on either side of the SDV aisle for the SDV to park.

3.2 Material handling assumptions

Various assumptions were required to model and simulate the cross-dock material handling activities of the two MHE configurations. We first discuss premises common to both models: the General Assumptions (GA), followed by those unique to each of the particular CD configurations.

- (GA 1) Trucks arrive at a CD with inter-arrival time 0, ready to offload or load (more correctly, the cross-dock never waits for a vehicle). This enables us to concentrate on the material handling tasks of the CD.

- (GA 2) The number of pallets carried on each IB truck is uniformly distributed between 20 and 26, that is, full capacity of a 53*ft* trailer. Those pallets are ready to offload.

- (GA 3) Pallets received from IB trucks are sorted, then loaded to respective outbound trucks as a *unit load* using MHE (forklifts or SDV).

- (GA 4) All outbound trucks (OB) require *UNIF* (20, 26) pallets to be loaded onto them, before leaving the CD.

- (GA 5) MHE are available throughout the simulation runtime, with no downtime.

- (GA 6) MHE are allocated based on their proximity to the "job call". By this, we mean the nearness of free MHE to the pallet in question.

- (GA 7) Each IB or OB truck has a changeover time[2] of *TRIA* (2, 3, 5) *mins*. That is the time taken for a processed truck to depart, or for an unprocessed truck to dock at a doorway.

Based on the offloading or loading activity in which it engages, a manual forklift is categorized as either an inbound or outbound forklift, respectively. An IB-FL, dedicated to offloading and staging pallets from an inbound truck to the respective staging area, can move across the inbound doors or between an inbound door and the staging area. An OB-FL sorts and loads pallets from the staging area to destination trucks at outbound doors, and can thus move between the staging area and those doors.

Modelling assumptions: forklift-only cross-dock facility (**Figure 2**)

1. An IB-FL assigned to a particular truck offloads pallets, one by one, from that inbound vehicle to the staging area. The time required depends on speed and travel distance.

2. The staging area ($58 \times 15 ft^2$) in front of an inbound door is reserved for pallets received from only that particular door.

3. An IB forklift assigned to offload an IB truck is relieved only after the offloading is complete.

---

[2] We are grateful to a warehousing consultant for this choice of distribution.

4. When the staging area is full, an OB-FL needs to move those pallets causing congestion. The assigned IB-FL will wait for that, before completing the offloading process.

Now, consider the *forklift-and-self-driving vehicle cross-dock* (**Figure 3**). The CD sortation and consolidation activities are the most cumbersome, resulting in huge floor and traffic congestion, and requiring immense labour and supervision to ensure transshipment accuracy. SDV are therefore particularly useful in combining several palletized loads. After the unloading of an inbound pallet, an SDV is programmed (with RFIDs or instructions from the master computers used for job allocation) to learn the destination of that pallet. The SDV will thereby transfer pallets to appropriate outbound doors, significantly reducing the labour requirement.

MHE in a forklift-and-SDV CD is categorized by the activities of offloading, sorting or loading, respectively performed by inbound forklifts, SDV and outbound forklifts. Pallets are offloaded from inbound trucks by IB-FLs, then are transferred to an SDV if available; otherwise, they go to the pallet holder. IB-FLs travel around or between inbound doors to offload the trucks. SDVs that are dedicated to sort and deliver pallets, from IB to OB doors, traverse the assigned 12 *ft* SDV aisle *via* a Manhattan-distance based path. A pallet that is on a pallet holder or on an IB-FL can be grasped by an SDV, but it can offload that pallet only to an OB-FL. As for OB-FL, they load pallets to outbound vehicles from SDV or from the staging area. An SDV sorts pallets and moves them to OB doors; OB-FL load pallets directly from the SDV to an outbound truck if available, or else places them in the staging area.

Our *modelling assumptions* for the forklift-and-self-driving vehicle cross-dock facility are as follows:

1. Pallets picked up by IB-FLs from inbound trucks are loaded to an SDV, if available at the designated SDV waiting zone, otherwise on the free pallet holder. When both are unobtainable, the inbound forklift awaits an SDV at the designated drop-off point.

2. After delivering the pallets, free SDV will return to the shortest active IB door (having the least number of SDV) within a 10-door distance.

3. Unavailable outbound vehicles cause pallets to be placed at the staging area to release the SDV. Sometimes there is no outbound truck and the staging area is full. There is then a delay in offloading, as the SDV anticipates arrival of an inbound vehicle.

4. Pallets staged are loaded by OB-FL to an OB truck, once latter is available at the door.

Simulation of the forklift-and-SDV CD proceeds *via* the general and modelling assumptions. **Table 1** presents particular design parameters and time delays.

3.3 Output performance measures

Only a few metrics provide insight on *overall* cross-dock performance; most of the others emphasize particular aspects of CD operation: MHE, doors or trucks. For example, output measures such as Average Throughput rate, Average Pallet Flow Time or CD Utilization rate offer intuition on overall effectiveness and general capability of a cross-docking operation. But metrics such as Outbound Truck Tardiness and Average Loading or Unloading time indicates the performance of only certain facets of that operation (e.g. truck scheduling procedure) in a CD. Recognition of suitable output measure(s) is thus important for proper evaluation of SDV in a cross-dock.

After discussion with CD subject matter experts, we chose *Average Throughput rate, Average Throughput rate/MHE* and *Average MHE Utilization rate*. Together, these measures will provide good intuition on CD material handling effectiveness and will enable assessment of SDV capability.

δ, the mean throughput rate, furnishes an understanding of the facility's overall material handling capability for a given MHE configuration.

Let P_j = Number of pallets processed on day j; N = Total number of replications or days.

$$\text{Then } \delta = \frac{\sum_j^N = i\ p_j}{N} \text{ in units of } \textit{pallets/day or ppd} \tag{1}$$

The cross-dock objective is to maximize δ, the mean throughput rate. However, efficient MHE utilization cannot be assured by studying only δ. Consider the ratio between total MHE uptime and the total time available in a CD, that is the *Overall* MHE Utilization rate, . This offers intuition on aggregate utilization of the CD's equipment, rather than attempting to analyse any machine individually. (Naturally, MHE idle time can incur cost in terms of non-ideal utilization of labour and MHE.) The category-wise MHE utilization rates are also computed for IB-FL (UI), SDV (US) and OB-FL (UOB). The CD objectives are to maximize the overall utilization, and each of those category rates.

Let UT_{ij} = Uptime of ith MHE on the jth replication; T = Simulation run time/ replication; c = Total number of MHE in a cross-dock; X_I =Total number of inbound forklifts in a CD; X_S= Total number of self-driving vehicles in a CD;

X_o = Total number of outbound forklifts in a CD.

With $c = X_I + X_O + X_S$ (and $X_S = 0$ for FL—only CD);

$$\text{Then } U_0 = \frac{\sum_i^{c=1}\sum_j^N UT_{ij}}{T \times c \times N} \times 100\% \tag{2}$$

Also, all for the jth replication:

Let UTI_{ij} = Uptime of ith IB-FL; UTS_{ij} = Uptime of ith SDV; and UTO_{ij} = Uptime of ith OB-FL.

$$\text{Then } U_I = \frac{\sum_i^{x=1}\sum_j^N UT_{ij}}{T \times X_I \times N} \times 100\% \tag{3}$$

$$U_S = \frac{\sum_{i=1}^{x_S}\sum_{j=1}^N UTS_i}{T \times X_S \times N} \times 100\% \tag{4}$$

$$U_{OB} = \frac{\sum_{i=1}^{x_0}\sum_{j=1}^N UTO_{ij}}{T \times X_O \times N} \times 100\% \tag{5}$$

While the MHE utilization rate specifies an awareness of equipment usage and idleness, it may not indicate *efficient* employment of MHE. Poor MHE allocation in a CD (between IB-FL and OB-FL, or between IB-FL, SDV and OB-FL) may still result in achieving a reasonable MHE utilization rate. That is, might a lesser amount of material handling equipment process a greater number of pallets? The following criterion, *Average Throughput Rate/MHE*, δ_M, seeks to resolve this.

δ_M attempts to compromise between the preceding two performance metrics. It is a ratio between Average Throughput rate (δ) and the total amount of MHE in the cross-dock. δ_M helps us understand how efficiently pallets are processed using the MHE in that CD. The cross-dock objective is to maximize

$$\delta_M = \frac{\sum_{i=1}^{N} P_i}{N \times c} \tag{6}$$

δ_M is measured in *pallets/day/MHE*, and c is the number of MHE.

3.4 ARENA simulation model overview

ARENA 15.0 was employed for CD modelling. For each of the forklift-only and forklift-and-SDV cross-docks, separate simulation models were developed. Those models followed the material handling assumptions and working parameters of Section 3.2. A Visual Basic API in ARENA was used to model a scalable facility, but (**Tables 1** and **2**) we fixed the size as 50 IB doors ×50 OB doors. Each respective model

Parameter	Value
Model naming	(IB-FL). (OB-FL)
CD working or simulation run time	7.5 *hr/day*.
Facility dimension	
Length	750 *ft*.
Width	97 *ft*.
Trucks	
Inbound or outbound truck inter-arrival time	0.
Pallets per inbound or outbound truck	*UNIF* (20, 26).
Docks	
No. inbound doors or docks	50.
No. Outbound doors or docks	50
OB staging capacity	13 Pallets in 15 × 13 ft^2
Number of pallet holders	1
Forklift average speed	
IB-FL or OB-FL with pallet	316 *ft/min* or 1.61 *m/s*.
IB-FL or OB-FL without pallet	548 *ft/min* or 2.78 *m/s*.
Time delays	
IB or OB truck changeover time	*TRIA*(2, 3, 5) *mins*.
IB or OB Forklift travel time	*manhattan Distance/Speed*
Forklift time to manoeuver and pickup pallets	*UNIF* (8, 12) *secs*.
Forklift time to manoeuver and drop pallets	*UNIF* (8, 12) *secs*.

Table 2.
Design parameters forklift-only cross-dock facility.

Factors	Notation	Number of levels	Factor levels
Number of IB-FL	X_I	5	25, 30, 35, 40, 45
Number of OB-FL	X_o	10	50, 60, 70, ..., 130, 140

Table 3.
RSM Factor Definition: forklift-only cross-dock. $5 \times 10 = 50$ *treatment combinations in full-factorial design.*

Factors	Notation	Number of levels	Factor levels
No. IB-FL No. SDV	X_I	5	25, 30, 35, 40, 45
No. OB-FL	X_S	11	50, 60, 70, ..., 140, 150
	X_O	7	25, 30, 35, 40, 45, 50, 55

Table 4.
RSM Factor Definition: Forklift-and-SDV cross-dock. $5 \times 11 \times 7 = 385$ *treatment combinations.*

would terminate when the simulation clock time reached 7.5 *hrs*, assuming one shift or 8 *hrs* of operation with 30*mins* break and no simulation warm-up period.

Validation of a simulation model is critical to ensure that the model represents the actual system [22]. Our models for the FL-only and FL-and-SDV cross-docks were validated by the group of members involving cross-dock SMEs and a simulation analyst from the SDV manufacturer.

Besides the input parameters in **Tables 1** and **2**, execution of the cross-dock simulation model requires the MHE configuration. For an FL-only CD, this includes X_I and X_O (**Table 3**); and those values plus X_S (**Table 4**) for an FL-and-SDV CD. These configurations determine the cross-dock's material handling performance, and (most of) the facility's variable operating cost.

Optimization of a simulation model, or statistical comparison of two models, requires a minimum of 30 replications each. The mean runtime for each replication of our ARENA CD simulation model took 1 minute, in a computer system with Intel(R) Core(TM)i7-3537U CPU @ 2.0 GHz processor, 6GB RAM. However, there are clearly very many feasible MHE configurations for the forklift-only CD and for the forklift-and-SDV cross-dock. To avoid the enormous computer run time, we propose a simulation-optimization technique in Sections 4 and 5 to overcome those computational difficulties. We suggest a *meta-modelling* approach to find optimal MHE configuration in a fast, time-efficient manner.

4. Response surface methodology

Fu [11] presents various simulation-optimization techniques, of which Response Surface Methodology (RSM) meets our requirement. The goal of our meta-modelling approach is to develop RSM prediction models for CD performance metrics using the MHE configuration as exploratory variables. We will then employ those regression prediction models to formulate constraints of an optimization problem.

Solution of the latter will yield the best MHE configuration, that is, one whose total variable operating cost is minimum, and whose cross-dock performance metrics are as desired.

A full-factorial experiment is designed for both CD simulation models with differing levels of MHE configuration. Performance of the FL-only CD is studied by varying the numbers of IB-FL and OB-FL in the CD. Those two factors have respectively 5 and

10 levels, as shown in **Table 3**. However, three factors (**Table 4**) are needed to evaluate the performance of the FL-and-SDV cross-dock: numbers of IB-FL, SDV and OB-FL in the CD. For both simulation models, each treatment combination is replicated[3] 15 times, recording the performance metrics (δ, δ_M, U_O, U_I, U_{OB}, U_S) in each case.

Four[4] regression models were estimated for each performance metric for both cross-dock facilities. Statistical assessment of those models showed that the regression coefficients of each model to be statistically significant. In light of R squared and the predicted regression sum of squares, we chose quadratic regression models. In view of the preceding statistical significance, further analysis enabled identification of the best MHE configurations.

Suppose R is a CD performance metric; its estimate is defined as $\hat{R}'$ for the FL-only CD, and as $\hat{R}''$ for the FL-and-SDV CD. The following remarks refer to both of those estimates.

As *quadratic* regression models, the fitted equations included terms linear in each of the variables, plus all possible products (every term of total exponent 2) of each variable multiplied by itself or any of the others. (Detailed regression equations are omitted for lack of space.)

5. Optimization of cross-dock MHE configuration

The fitted regression models for each facility's performance metrics were used to formulate respective optimization models. Two independent minimization models were formulated in terms of MHE configuration, for the FL-only CD and for the FL-and-SDV CD. Each cost objective was subject to constraints by its predicted performance metrics $\hat{R}$ that yielded the desired level of performance R. Each optimization model thus furnished an optimal or near-optimal[5] MHE configuration and the corresponding total variable cost for the CD.

5.1 MHE optimization models

CD variable costs include labour and MHE operating costs. Each forklift truck naturally requires an operator; we assume that no additional labour is needed for material handling in a cross-dock. The total variable cost (TVC) of operating an FL-only CD is estimated to be $\$522.63/day$. (Figure includes operator wages, and forklift power consumption and repair and maintenance.)

Decision Variables $\boldsymbol{X_I}$ - Number of IB-FLs; $\boldsymbol{X_O}$ - Number of OB-FLs.

The resulting optimization model for an FL-only CD is then formulated as:

$$\textbf{Min } \text{TVC} = 522.63(X_I + X_O) \tag{7}$$

Subject To $\widehat{\boldsymbol{\delta}'} \geq \bar{\delta}$; $\widehat{\boldsymbol{\delta}'_M} \geq 200$; $\widehat{\boldsymbol{U}'_O} \geq 80\%$; $\widehat{\boldsymbol{U}'_I} \geq 60\%$; $\widehat{\boldsymbol{U}'_{OB}} \geq 60\%$;
And $25 \leq \boldsymbol{X_I} \leq 45$; $50 \leq \boldsymbol{X_O} \leq 140$; $\boldsymbol{X_I}, \boldsymbol{X_O} \in \boldsymbol{Z}^+$ (*positive integers*).

---

[3] Chosen following the number of replications (= 10) used by Shi et al. [13] to fit a cross-dock response surface model, under similar situation.

[4] Regression model with linear terms, linear and interaction terms, square terms and square and interaction terms, respectively.

[5] Based on predictive accuracy of the regression models.

Manual forklifts in a CD incur the costs of labour and of forklift operation; just operating costs are incurred by SDV. Provided that, over time, the manual forklifts and SDV have about the same operating costs, the following optimization model will minimize the total variable cost of a forklift-and-SDV cross-dock.

Decision Variables: X_I - Number of IB-FL; X_S - Number of SDVs; X_O - Number of OB-FLs.

$$\textbf{Min}\ \mathrm{TVC} = 522.63(X_\mathrm{I} + X_O) + 271.20X_S \tag{8}$$

Subject To: $\widehat{\delta'} \geq \bar{\delta}$; $\widehat{\delta''_M} \geq 120$; $\widehat{U''_O} \geq 80\%$; $\widehat{U''_I} \geq 60\%$; $\widehat{U''_{OB}} \geq 60\%$; $\widehat{U''_S} \geq 60\%$
And $25 \leq X_I \leq 45$; $50 \leq X_s \leq 155$; $25 \leq X_O \leq 55$; $X_I, X_S, X_O \in Z^+$.

(The coefficient 271.20 of X_S is the estimated average cost/day to operate SDV including electricity, repair and maintenance).

5.2 Optimal MHE configurations

The optimization models for the respective cross-docks are mixed-integer nonlinear programmes (MINLP). Each objective function is linear, with nonlinear constraints and integer decision variables. Each MINLP model is solved using the Lingo 17.0 solver for an increasing level of desired throughput rate $(\bar{\delta})$.$\bar{\delta}_M$ is set as 200 *ppd/MHE* for the FL-only CD, and as 120 *ppd/MHE* for FL-and-SDV CD.

The latter cross-dock naturally requires a greater number of MHE, to achieve a δ similar to the former. Optimal MHE configurations (solutions of MINLP models) are in **Table 5**.

The operating costs for Forklifts and SDVs, at the time of this research, were assumed at extreme limits to account for maximum cost. Adjusting to industrial standards, the average cost to operate a forklift drops to \$348.70/day and \$97.28/day for SDVs. The revised objective functions for the cross-docks are presented in Eqs. (9) and (10). The corresponding optimal MHE configurations, total variable costs and payback period (PP)[6] are found in **Table 6**. (Optimal configurations for the FL-only

$\bar{\delta}/10^3$	FL-only		FL-and-SDV			TVC (\$/day)		Optimal model
	X_I	X_O	X_I	X_S	X_O	FL-only	FL-and-SDV	
20	32	60	33	81	31	48,082	55,416	FL-only
22	33	67	35	90	34	52,263	60,469	FL-only
24	34	74	36	98	39	56,444	65,775	FL-only
26	36	81	38	110	41	61,148	71,120	FL-only
28	38	89	39	115	52	66,374	78,747	FL-only
30	42	97	41	127	55	72,646	84,615	FL-only

Table 5.
*Optimal MHE configuration, if FL operating cost = \$522.63/*day, *SDV operating cost M = \$271.20/*day.

[6] No of days in a year = 5 days × 4 weeks × 12 months; one 8 hr. shift per day. Cost of a Forklift = \$ 75,000. (Source: http://www.costowl.com/b2b/forklift-electric-cost.html).

DOI: http://dx.doi.org/10.5772/intechopen.1005502

$\bar{\delta}/10^3$	FL-only CD			FL-and-SDV				Optimal model	Savings/ year ($)	Fixed cost (in $M)	PP[1] (in years)
	X_I	X_O	TVC ($)	X_I	X_S	X_O	TVC ($)				
20	32	60	32,080	32	96	26	29,563	FL-and-SDV	604,061	17.1	14.15
22	33	67	34,870	33	102	31	32,239	FL-and-SDV	631,354	17.93	14.2
24	34	74	37,660	36	112	33	34,956	FL-and-SDV	648,946	19.8	15.26
26	36	81	40,798	38	117	38	37,883	FL-and-SDV	699,586	20.4	14.58
28	38	89	44,285	40	130	41	40,891	FL-and-SDV	814,512	22.73	13.95
30	42	97	48,469	41	127	55	45,830	FL-and-SDV	633,490	21.23	16.75

[1]*No of days in a year = 5 days × 4 weeks × 12 months; one 8 hr. shift per day. Cost of a Forklift = $ 75,000. (Source: http://www.costowl.com/b2b/forklift-electric-cost.html).*

Table 6.
*Optimal MHE configuration (FL operating cost = $348.70/*day; *SDV operating cost = $ 97.28/ day). Payback period is time required to recoup extra fixed cost*[1] *of SDV in FL-and-SDV CD, relative to FL-only CD, for that* δ.

CD, as shown above in **Table 5**, still hold; the optimal MHE arrangement changes only for the FL-and-SDV CD.)

$$\text{For FL} - \text{only CD} : \mathbf{Min}\ 348.70\ (X_I + X_O) \tag{9}$$

$$\text{For FL} - \text{and} - \text{SDV CD} : \mathbf{Min}\ 348.70\ (X_i + X_O) + 97.28\, X_S \tag{10}$$

5.3 Validation of optimal MHE configuration

Suppose the regression model predictions were accurate (i.e. $\hat{\delta}$ is not statistically different from δ). MHE configurations suggested by solving the MINLP model for FL-only and FL-and-SDV CDs would then be *optimal*, and would furnish performance measures equivalent to the predicted performance. Alternatively, prediction errors may yield only "near-optimal" MHE configurations. As mentioned earlier, there is a non-random source of variation, left unexplained by the quadratic regression models for the CD performance metrics $\hat{R}'$ or $\hat{R}''$. Consequently, those fitted models do not predict CD performance fully accurately.

The solution of the MINLP models furnished "the best" MHE configurations, which were input into the two respective cross-dock simulation models. Could those CDs, that is the FL-only and FL-and-SDV cross-docks, operating under the configurations determined, furnish throughput rates that are similar? Statistical comparisons of mean throughput rates for all pairs of 'proposed optimal MHE configurations' are given in **Table 7**. That table shows *dissimilar* respective performance metrics (i.e. statistically significant differences between the mean throughput rates of the FL-only and FL-and-SDV CD).

We conclude, therefore, that by solving the MINLP models, MHE configurations obtained are efficiently allocated. Although not necessarily the optimal ones, those configurations are near-optimal.

6. Discussion and suggestions for further research

Solutions of the formulated MINLP model yield many interesting insights:

$\bar{\delta}/10^3$	FL-only				FL-and-SDV				t-Test p-value	
	Model	$\hat{\delta}'$	δ'	**TVC ($)**	**Model**	$\hat{\delta}''$	δ''	**TVC**	**n = 8**	**n = 30**
Operating cost for FL = $522.63/day and SDV = $271.20/day										
20	32.60	20,105	19,512	48,082	33.81.31	20,001	19,629	55,416	0.03	0.00
22	33.67	22,162	21,681	52,263	35.90.34	22,021	21,793	60,469	0.02	0.01
24	34.74	24,049	23,988	56,444	36.98.39	24,000	19,056	65,775	0.00	0.00
26	36.81	25,989	26,244	61,148	38.110.41	26,002	24,928	71,120	0.00	0.00
28	38.89	27,956	28,820	66,374	39.115.52	28,027	27,268	78,747	0.00	0.00
30	42.97	30,039	31,486	72,646	41.127.55	30,033	30,431	84,615	0.00	0.00
Operating cost for FL = $348.70/day and SDV = $97.28/day										
20	32.60	20,105	19,512	32,080	32.96.26	20,029	20,524	29,563	0.00	0.00
22	33.67	22,162	21,681	34,870	33.102.31	22,023	21,376	32,239	0.00	0.00
24	34.74	24,049	23,988	37,660	36.112.33	24,013	23,485	34,956	0.00	0.00
26	36.81	25,989	26,244	40,798	38.117.38	26,012	24,808	37,883	0.00	0.00
28	38.89	27,956	28,820	44,285	40.130.41	28,009	27,397	40,891	0.00	0.00
30	42.97	30,039	31,486	48,469	41.127.55	30,033	30,431	45,830	0.00	0.00

Table 7.
Statistical validation of optimal MHE configuration (for model "naming convention," see rows of ***Tables*** *5 and* ***6*** *for corresponding $\bar{\delta}$).*

- Total numbers of MHE in each category almost always increase together for an increase in desired throughput $\bar{\delta}$ (Sole exception: last two rows of **Table 7**, for FL-and-SDV CD)

- Statistical comparisons of manual and semi-automated CDs, under optimal MHE configurations (**Tables 5** and **6**), show significant differences in actual performances. For reduced MHE operating costs, a mixture of forklifts and SDVs in a CD results in lower total variable operating cost than for a purely-forklift facility.

The proposed solution methodology to find the optimal MHE configuration for FL-only and FL-and-SDV CDs operating at similar performance levels yields an *efficient* MHE arrangement.

Our solution to the MINLP model may be sub-optimal, but visual comparison (**Table 8**) of the manual and semi-automated cross-docks yielding similar δ show it could be financially beneficial to choose FL-and-SDV over the FL-only CD. A 5% confidence interval (CI) was constructed for the difference in mean throughput rates of the two CD types. That CI contained the value 0, leading to the conclusion that the two facility types had similar output performance δ. Savings/year and payback period[7] (PP), given in that table, support those possible benefits. (Note that **Table 8** assumes *two* shifts/day, whereas all other tables assume a single shift).

[7] No of working hours = 8 × 2 × 5 × 4 × 12 (two shifts per day); Operating cost, M = $ 97.28/day.

FL-only CD			FL-and-SDV CD			CI for difference in mean			Savings/ year ($)[1]	PP (in years)[1]
Model	δ'	TVC ($)	Model	δ''	TVC ($)	Lower bound	Mean	Upper bound		
32.72	23,412	36,265	36.112.33	23,485	34,956	−147	−73	1	628,387	22.56
36.81	26,244	40,798	39.125.39	26,260	39,359	−71	−17	38	690,864	19.65
39.94	30,441	46,377	41.127.55	30,431	45,830	−48	9	67	262,723	61.95

[1]*No of working hours = 8 × 2 × 5 × 4 × 12 (two shifts per day); Operating cost, M = $ 97.28/day.*

Table 8.
MHE configurations of cross-docks with similar δ.

For future research, we propose the following: The predictive accuracy of the regression models might be improved by incorporating additional explanatory variables (e.g. product mix, or labour activities other than for forklift operators); the MHE configuration computed for each CD would then be closer to the true optimum. An expanded MINLP cross-dock model could include additional constraints on costs or space. Finally, the model could be enhanced by considering further randomness in supply and demand, MHE breakdowns, CD traffic congestion and LTL shipment operations.

6.1 Pros and cons of proposed solution methodology

Among the positive features, the methodology of this chapter may in fact be used to solve complex simulation-optimization problems of any system, not just CDs, with a larger search space in a shorter time span. Moreover, multiple systems could be optimized and compared, subject to constraints on various performance metrics, eliminating family-wise error rate due to multiple comparisons.

On the negative side, in-depth knowledge of the system is required to model precisely, and to identify the factors which impact the overall performance of the system. The results could be near-optimal, but may be sub-optimal, if the regression models do not adequately address the sources of variation.

Though this solution methodology is very versatile and could be adapted to optimize and compare any other system, we caution future researchers that the need for in-depth understanding of the actual system is paramount when it comes to simulation modelling. Otherwise, the simulation model risks misrepresenting the system with distorted results. This risk could be mitigated by:

a. Involving subject matter expert(s) from the system and other stakeholders closely in the design, verification and validation of the modelling process.

b. Cross validate the model's functions and performances against the system.

c. Iteratively addressing the design gaps until the simulation model adequately represents the defined functional scope of the system.

Funding Statement

Research is supported by the Natural Sciences and Engineering Research Council of Canada, Grant # RGPIN-2019-06207.

Author details

Saravanan Natarajan and James H. Bookbinder*
University of Waterloo, Waterloo, Canada

*Address all correspondence to: jbookbinder@uwaterloo.ca

IntechOpen

References

[1] Natarajan S. Optimization and Comparison of Manual and Semi-Automated Material Handling in a Cross-Dock Using Discrete-Event Simulation. [Master's Thesis]. Canada: University of Waterloo; 2018

[2] Bartholdi JJ, Gue KR. Reducing labor costs in a less-than-truckload crossdocking terminal. Operations Research. 2000;**48**(6):823-832

[3] Gelareh S, Glover F, Guemri O, Hanafi S, Nduwayo P, Todosijević R. A comparative study of formulations for a cross-dock door assignment problem. Omega. 2020;**91**:102015

[4] Wolff P, Emde S, Pfohl HC. Internal resource requirements: The better performance metric for truck scheduling? Omega. 2021;**103**: 102431

[5] Boysen N. Truck scheduling at zero-inventory cross docking terminals. Computers & Operations Research. 2010;**37**(1):32-41

[6] Magableh GM, Rossetti MD, Mason S. Modeling and analysis of a generic cross-docking facility. In: Proceedings of the Winter Simulation Conference. 2005. 8 pages. Available from: https://informs-sim.org

[7] Liu Y, Takakuwa S. Simulation-based personnel planning for materials handling at a cross-docking center under retail distribution environment. In: Proceedings of the 2009 Winter Simulation Conference (WSC). 2009. pp. 2414-2425. Available from: https://informs-sim.org

[8] Yang KK, Balakrishnan J, Cheng CH. An analysis of factors affecting cross docking operations. Journal of Business Logistics. 2010;**31**(1):121-148

[9] Adewunmi A, Aickelin U. Optimisation of a crossdocking distribution Centre simulation model. SSRN Electronic Journal. 2010. DOI: 10.48550/arXiv.1003.37

[10] Torbali B, Alpan G. A literature review on robust and real-time models for cross-docking. International Journal of Production Research. 2022:1-30. DOI: 10.1080/00207543.2022.2062580

[11] Fu MC. Handbook of Simulation Optimization. New York, NY: Springer Publishing Company; 2014

[12] Aickelin U, Adewunmi A. Simulation optimization of the crossdock door assignment problem. SSRN Electronic Journal. 2008. DOI: 10.48550/arXiv.0803.1576

[13] Shi W, Liu Z, Shang J, Cui Y. Multi-criteria robust design of a jit-based cross-docking distribution center for an auto parts supply chain. European Journal of Operational Research. 2013;**229**(3): 695-706

[14] Suh ES. Cross-docking assessment and optimization using multi-agent co-simulation: A case study. Flexible Services and Manufacturing Journal. 2015;**27**(1):115-133

[15] Peixoto R, Dias L, Carvalho MS, Pereira G, Geraldes CAS. An automated warehouse design validation using discrete simulation. In: 2016 IEFE 19th International Conference on Intelligent Transportation Systems (ITSC). 2016. pp. 199-204

[16] He X, Prabhu VV. Design and analysis of AGV-based cross-docking

operations using analytical models. Production & Manufacturing Research. 2022;**10**(1):428-449

[17] Rupp J, Boysen N, Briskorn D. Optimizing consolidation processes in hubs: The hub- arrival- departure problem. European Journal of Operational Research. 2022;**298**(3):1051-1066

[18] Guo P, Weidinger F, Boysen N. Parallel machine scheduling with job synchronization to enable efficient material flows in hub terminals. Omega. 2019;**89**:110-121

[19] Sayed SI, Contreras I, Diaz JA, Luna DE. Integrated cross-dock door assignment and truck scheduling with handling times. TOP. 2020;**28**(3): 705-727. (Springer)

[20] Fragapane G, De Koster R, Sgarbossa F, Strandhagen JO. Planning and control of autonomous mobile robots for intralogistics: Literature review and research agenda. European Journal of Operational Research. 2021;**294**(2): 405-426

[21] Bartholdi JJ, Gue KR. The best shape for a crossdock. Transportation Science. 2004;**38**(2):235-244

[22] Kelton WD, Sadowski R, Zupick N. Simulation with Arena. Fifth ed. New York, NY: McGraw Hill; 2014

Chapter 3

Forces Transforming Transport and Logistics into Smarter Sustainable Systems

Felix-Eduardo Bueno-Pascual

Abstract

Markets complexity has grown to some extent that modern transport and logistics (T&L) is difficult to understand what to focus on, being impacted by the global economy, compliance with CO_2 emissions, competition, reorientation of business models, financial, and technology change, among others. So, it is important to identify what forces are transforming the T&L industry, which can be classified into five major pillars: Digitalization, Shifts in international trade, Enterprise-software-driven process changes, Changes in the market's domestic commerce, and Machine-driven process changes. These forces are visible and impactful; thus, businesses should be reshaped by anticipating disruptions regarding changes in distribution channels, economic growth outlooks, and the impact of technology changes. These forces will make logistics networks change as much as their environments change due to climate change and CO_2 emissions, using co-opetition to increase efficiency and adapt their business by building smarter sustainable systems to meet future challenges. Moreover, through three cases of study, it is seen how these trends are helping businesses not only to improve their operations, increasing performance, efficiency, and effectiveness, as well as to gain competitive advantage, but also to reduce its impact related to the climate change in compliance with Sustainable Development Goals (SDGs) and Environmental, Social and Governance (ESG) practices.

Keywords: shifts in logistics markets, digital impact on transport logistics, logistics driving trends, logistics digitalization, logistics transforming forces

1. Introduction

Companies and industries are in the middle of transformation, marked by huge changes in technology, with high disruption and competitive fervor to satisfy consumer preferences while they are dealing with barriers to entry and the optimization of supply chain and logistics. These changes can fit under the industry 4.0 umbrella to achieve the strategy of digital integration of end-to-end value chain (**Figure 1**), including development of digital business models, corporate culture, management approaches, role of IT, digitization of product and services offerings, and adopting data and analytics as core capabilities. All these changes cause complexity in business,

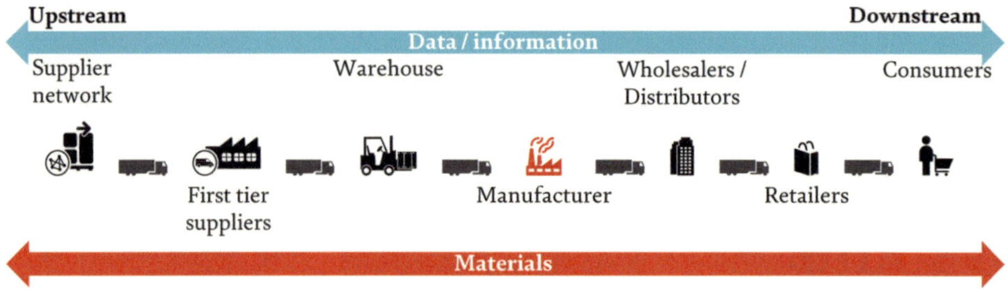

Figure 1.
End-to-end value chain.

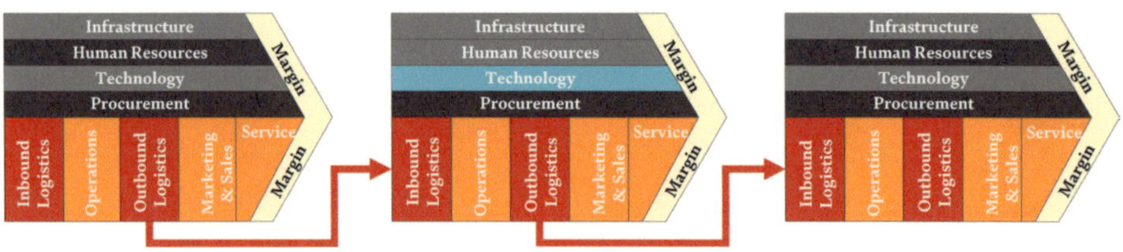

Figure 2.
Connected value chains.

and they are related to Chief Executive Officers (CEOs) main concerns aligned with sustainability [1, 2]:

- Revenue growth through company's prospects.
- Business disruptions due to changes in distribution channels.
- Business disruptions due to changes in core technologies of service provision.
- Availability of digital skills in the workforce and the industry.
- Improvement of global economic growth.

In terms of transformation as a driving force, innovative technologies are targeting logistics process to increase efficiency and productivity through automated scheduling, consolidation, on-demand trucking, and carrier-based analytics, which impact transportation industry since they must increase visibility across the supply chain.

Transportation is an important part of the supply chain, since it is the link between companies or business units by delivering materials from one location to another, being part of the end-to-end-value chain (**Figure 1**) through the inbound and outbound logistics functions from Porter's value chain with primary and support activities [3].

Transportations and logistics are important activities across world, since they represent about 10.7% of the Global Gross Domestic Product [4], which includes cargo transportation services and represent one of the backbones of international trade; it is being transformed by changing nature of digitalization and economic patterns.

Every company has its own way of managing its processes, but the general structure is as shown in **Figure 2**, with primary activities (inbound logistics, operations, outbound logistics, marketing and sales, and service) and support activities (infrastructure, human resources, technology, and procurement). Companies are connecting their value chains through inbound and outbound logistics.

Since the world is moving into an energy-constrained and low-carbon, supply chains need to ensure the cost of emissions is paid by those who reap benefits spurred by consumer behaviors and regulations, using technology that enables real-time control and greater flexibility. To improve productivity and achieve high performance, transport and logistics require aligned business strategy, agility, coordination integration, transparency, engagement, latest tools adoption, strengthening shortcomings, improvement of the cash flows, climate change impact considerations, as well as the management of customer expectations. These requirements can be met with high efficiency and effectiveness through forces transforming transport and logistics into smarter sustainable systems, such as:

- Global trade.
- Market's domestic commerce.
- Digitalization.
- Machine-driven processes.
- Software-driven processes.

2. Background and context

2.1 Transportation and logistics overview

Logistics is a strategic process connecting supply chain network through the management of information and physical materials, involving the movement and storage of materials, parts, and finished inventory across organization to fulfill customer orders [5–7].

Logistics as process involves six main functions to deliver value through flow of information and materials (**Figure 3**). These functions are divided into six components [8]: logistics will acquire raw materials and supply to manufacture or deliver

Figure 3.
Logistics functions.

products or services (procurement), will store and manage inventory (inventory management) in a distribution centre or warehouse (warehousing), it will identify materials used to protect goods while moving across locations (packaging) and, it will also move goods from a source to a destination point (transportation) to satisfy received customer orders (order fulfillment); also, for each component it is required information to manage these operations and communicate with customers (information management).

Transport and logistics (T&L) is one of the most important industries worldwide, represented by an umbrella of vehicles and road traffic, rail transport, aviation, water transport, logistics, and public transport and mobility services [3–9]. With the COVID-19 pandemic outbreak, the Suez Canal blockage, and the Russia–Ukraine war, all industries were affected, as well as T&L marked, increasing the imbalance between supply and demand due to supply shortages and the increase in price pressures.

Freight transport and logistics rates were decreasing, impacting global container freight rates, which started to decrease in January 2022, while global air freight rates increased from China to North America and from China to Europe, remaining stable from Europe to North America [10, 11].

On the other hand, geopolitical disruptions also impacted on different industries through commodity prices, increasing prices for goods, metals, and energy, the supply chain for manufacturing goods was impacted, causing bottlenecks and delays, and trade flow uncertainty increased [9–11].

Supply chain disruptions have raised challenges for companies such as [6–12]:

- Labour constrains and capacity.
- Need for increased capacity.
- Shipments delay.
- Reduction of long-dwelling containers.
- Decrease of deal value.
- Decrease of transaction activity.
- Increase of freight rail services.
- Increase of pricing.

Due to challenges, companies need to increase flexibility, efficiency, and productivity while reducing costs to be effective. From 2015 to 2050, it is expected a global trade growth in volume by four times, while the value of goods transported increased up to US$68.5 trillion [12], for which it is necessary to identify drivers impacting the transport and logistics industry for the next years, such as global uncertainty, digitalization, innovation in technology, talent, and new skills required, costs, transparency, and sustainability, as well as global uncertainty, causing disruption in supply chain and business models. These drivers will be identified across the industry and best practices through a set of trends, which are addressed in the next section.

DOI: http://dx.doi.org/10.5772/intechopen.1005001

2.2 Transport and logistics over time

Transport and logistics have transformed over time, and big changes can be tracked across big globalization waves, as shown in **Table 1**.

Transport and logistics industry has crossed across time, enabled by globalization, connecting countries and economies to improve people's welfare through products and services; however, due to climate change, there are new technologies changing

Wave	Logistics	Transport
First wave (Manufacturing Era)	• Export of goods. • Opening of new markets. • Increase of manufacturing exports.	• Cost reduction. • Bulk ships and liners. • Use of steamships and trains as a main transport mode.
Second wave (Communication Era)	• Lean thinking. • Just-In-Time and Kanban. • Postponement. • Marketing Mix: Price, Product, Place, and Promotion. • Integrated Logistics Management.	• Car and planes were born. • Use of trucks. • Free trade vehicles. • Increase of physical distribution importance. • Industrial tariffs. • Limited scale shipping. • High-speed rail.
Third wave (Technology Era)	• Cross-Docking. • Quick Response. • Outsourcing (BPO) and Co-Sourcing. • Agility. • Radio-Frequency Identification (RFID). • Computer-Aided Dispatching. • Collaborative Planning, Forecasting and Replenishment (CPFR). • Vendor Managed Inventory. • E-commerce and internet. • SCM enterprise-wide integration.	• Geographic Information Systems (GIS). • Centralized Tariff Control (CTC). • Large-scale air cargo. • Transport networks. • Electric and Hybrid cars.
Fourth wave (Digital and Information Economy Era)	• Lead by digital practices • Industry 4.0 • Nearshoring practices • Workforce collaboration • Responsible datafication • Blockchain • Internet of things • Distributed infrastructure • Digital factories	• Global effect of climate change. • Clean energy and sustainable technology. • Self-driving vehicles. • Drones.

Table 1.
Transport and logistics across globalization waves [6–10, 18].

and reshaping the world, as shown in **Table 1**, which are part of the forces transforming transport and logistics into sustainable systems.

2.3 Climate change and sustainable development goals

Climate change is leading businesses through sustainable practices because of long-term shifts in weather patterns and temperatures. Moreover, according to the 27th Annual CEO Survey from PwC, many companies are looking to reinvent themselves, and they believe that they would not be viable in 10 years if they continue moving across the same path due to technology, change in customer preferences, government regulations, competitor actions, supply chain instability, and climate change; moreover, according to PwC, among 55% of the global Gross Domestic Product (GDP) is moderately or highly dependent on nature [8]. Among the initiatives they are focusing to face these challenges are [8–10]:

2.3.1 Climate risk, adaptation, and resilience

Looking for different ways to mitigate climate risk, while increasing business resilience, identifying and focusing on the right outcomes to create competitive advantage with compliance and long-term value.

2.3.2 Sustainable development goals (SDGs)

Goals adopted by United Nations Member States in 2015, which are working in partnerships to create peace and prosperity for people and the planet. In terms of business, they need to align with the SDGs to help countries achieve their goals through the supply chain impact, future strategy impact, awareness, prioritization, measurement (SDG Strategy Analytics), and reporting.

2.3.3 Energy solutions

Business are working to reduce Green House Gas (GHG) to reach a net zero compliant economy, changing the way they use energy, prioritization of decarbonization, digitization of processes, reducing emissions and costs through efficiency and energy saving, and mitigating risks while creating value.

2.3.4 Net zero transformation

Since every business and industry is different, they are tailoring and targeting solutions to reduce carbon emissions by starting from strategic points, transforming the business, assessing the value chain and decarbonization, as well as increasing transparency in legal and regulatory compliance.

2.3.5 Investing with impact

Every change implies challenges, and this is no exception, so they need to consider the Environmental, Social, and Governance (ESG) impact when investing, trying to mitigate risks and reduce negative outcomes for people, planned, stakeholders, and contribute with positive change. Since investors, shareholders, and other stakeholders are expecting businesses to pursue sustainability, companies are focusing on implementing strategies aligned to green taxes and incentive trackers, building trust through transparency and holistic thinking, sustainability assurance, and reporting.

In the last SDG Report from 2023, the United Nations stated that climate action (Goal 13) requires taking urgent intervention to mitigate climate change and its impacts, and GHG emissions need to be reduced in all sectors, which requires [11]:

- Goal 6 Clean water and sanitation: Sector-wide investment and capacity-building need to be increased since there is water stress and scarcity.

- Goal 7 Affordable and clean energy: Renewable energy needs to be done because currently, renewable sources power nearly 30% of energy consumption.

- Goal 9 Industry innovation and infrastructure: Reduce global carbon dioxide emissions from energy combustion and industrial processes.

- Goal 12 Responsible Consumption and Production: Implement policies that support shifts to sustainable practices that allow the decoupling of economic growth from the use of resources.

- Goal 17 Partnerships for the Goals: Developing countries are requiring necessary financing and technologies to accelerate the SDGs implementation.

Even though there are 17 SDGs, there were highlighted those related with Climate Change, but all of them are important and need to be addressed.

3. Methodology

The world is changing, and businesses and the way they manage their value chain are being impacted by technology, globalization, climate change, competence, and international and local regulations. So, this research was carried out by reviewing three main sources: Literature, industry best practices, and case studies (**Figure 4**).

Figure 4.
Research methodology.

Each component of the analysis was made up as follows:

- Literature Review: The understanding of logistics and transport require the knowledge of what it is, how it is done, and what are some current trends driving those changes, such as the climate change and the sustainable development goals (SDG), which were considered to analyze the trends moving logistics and transport.

- Best practices: Every industry is currently led by some trends, and transport and logistics is no exception, because of which, as part of them, it is included as part of the results best practices aligned to digitalization, business transformation, transport and logistics, industry, and logistics 4.0.

- Case studies: Finally, by analyzing the literature and the best practices, some case studies were reviewed as part of own experience in projects related to logistics transformation, logistics and supply chain, and logistics and digitalization, so it can be seen how all trends are reshaping the industry.

4. Transportation and logistics trends

Based on the global market and the issues for the industry, there are some important trends in transport and logistics industry, which are [11–14]:

- Sustainable funding mechanisms and transparency.
- Adoption of electric vehicles.
- Transportation systems modernization.
- Resilient transportation networks.
- Digital and technological innovation.
- Use of blockchain platforms.
- Global uncertainty.
- Talent new skills.
- Growth of e-commerce.
- Green 3PL.
- Changes in regulations, laws, policies, and tax regulations.
- Market risk from interest and exchange rates.
- Fraud and corruption.

- Emerging transportation concepts.
- Slowdown in economy.

Trends can be seen from the more strategic impact to the most operative, and they can be classified as shown in **Figure 5**.

4.1 Global trade

International trade is impacting highly on all markets and industries, in land transports from China to the European Union (EU), and it is expected to be intensified due to many factors such as pandemic results, war, political stability, and overall leading economies. There is expected growth in the transport corridors between China and the EU through the Road and Belt Initiative and other emerging economies in the next few years, which will lead to lower transportation costs and the creation of new services. Among the opportunities that are being leveraged by this trend are:

- Trade cost and investment reduction can be achieved by using emerging trade routes such as the Belt and Road Initiative corridors between China and Europe.
- Infrastructure modernization and hubs located across main transport corridors.
- Increase accessibility to new business areas due to low logistics costs.
- New trade agreements across routes.
- Emerging market trade flows that allow scaled services offerings.

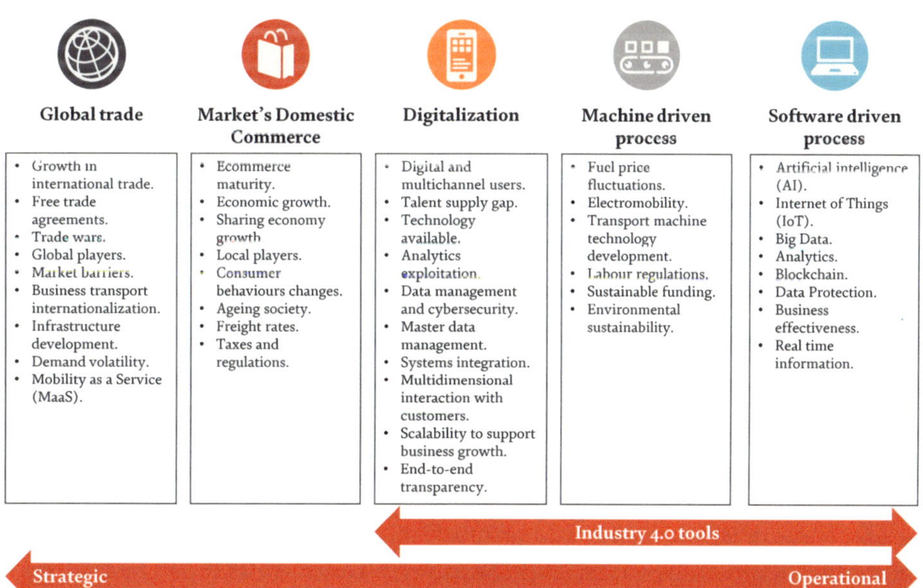

Figure 5.
Transport and logistics trends.

- Supply chain strategy adjustment to benefit from delivery time and cost reduction.

- Trends leveraged by companies are mainly related to [15]:

- Transportation infrastructure investments: To improve efficiency and effectiveness of logistics processes, reduce transit times and reduce, as a consequence, delivery time.

- New cross-border services: To reduce costs inside countries to impact overall costs of delivery. Companies are looking to leverage transport fleets with the increasing volumes between China and EU.

4.2 Market's domestic commerce

Domestic commerce involves within-border transactions of a company, either sales or buying of goods and/or services that can be delivered online or physically [16]. Domestic commerce for companies is increasing across regions, through various levels of optimization, transportation, and logistics, which helps them to create a shared economy and value-chain integrations between the focal firm, the producers, the consumers, and Third-Party Logistics (3PL) companies. However, the first step of integration is to comply with local regulations, satisfy customers' needs and have capabilities to compete with local market players. Hence, it is important to leverage the use of best practices like e-commerce to plan and create strategic alliances to satisfy the market with low costs. Among the solutions used as part of the market's domestic commerce trends are [17–20]:

- E-commerce in big business [17]: Big companies are using e-commerce to reduce costs and time to market and integrate the supply chain while increasing profits. Some big companies are looking towards opportunities to offer their products online through different product categories such as Incubation (products with low or medium levels of growth and occurrence of exceptional periods, such as groceries and health and beauty), growth (products that can increase growth rate significantly, such as fashion, electronics, household equipment, and furniture and decoration), slowdown (those whose growth rate is below the level achieved in growth phase, such as e-books and travelling), and saturation (for goods with very low level of growth and high maturity of online channel).

- E-commerce used in logistics [18]: Companies are investing in e-commerce to manage customer orders and suppliers' deliveries, connecting processes to manage inbound logistics, production, outbound logistics, and marketing and sales, reducing costs while improving the company's performance and operations. It is expected that e-commerce businesses will invest in logistics in the longer term to seek possibilities to close value chains.

- Courier, Express and Parcel [19]: The use of 3PL companies (CEP) focused on the 80% of customers with 20% of the revenue helps to manage the 20% of customers with 80% of revenue, increasing the value performance, efficiency, and effectiveness to compete in local markets. CEP companies will need to design tailor-made solutions for e-commerce over longer-term due to omnichannel sales, taking into account factors like delivering options offered, ability to pick up at a location, return policies, delivery speed, retailer reputation, product selection, consumer/peer reviews, and detailed product information.

- Economy solutions [20]: Use of collaborative logistics to optimize the management of transportation, warehousing, and supply chain with other companies to reduce costs and respond to customer pressures. An example of this practice is the consolidation of logistics. The sharing economy is being implemented in supply chain management, road transport, and freight forwarding.

4.3 Digitalization

Digitalization is a substantial change that is changing all industries, and it is expected to reshape businesses. However, it is important to have a clear definition of digitalization, which can be defined as a framework used to exploit digital opportunities through a transformation process – used to restructure economies, institutions, and society on a system level – and technologies – such as big data, sensors, 3D printing, etc. – to create, propose and capture value in the business to be offered to future customers [21]. Digitalization offers value, allowing business to adapt to changes through resilience, dynamic capability, transition models, and increased competitiveness in the market. To maximize digital value, it is important to consider the elements in **Figure 6**, which will help businesses to be prepared to take advantage of opportunities like:

- Simplified internal process.
- Digital reach to final customers.
- Online sales and marketing.
- Talent supply gaps.
- Cost reduction.
- Risk reduction (such as risk related to online payments).
- Client needs addressing.
- Increased revenues.

Elements in **Figure 6**, are related to:

- Business vision [22]: Focus on what business want to achieve to increase integration, such as value-chain integration, customer access, sales, channels and marketing, vertical value-chain integration, overall digitalization, product development and engineering, digital business models, and product service portfolio. It will provide new business models, transaction types, marketplaces, and services offering as revenue sources, since digitalization is not only the implementation of information and communication technologies but also a framework to add value to business and impact customers.

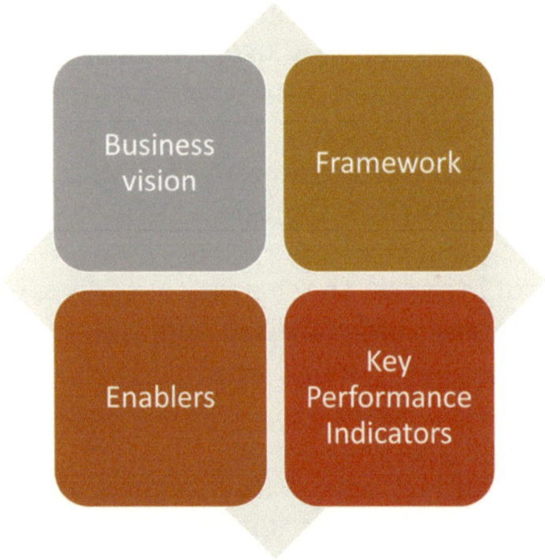

Figure 6.
Digitalization components.

- Framework [23]: As part of a digital transformation and digitalization, it is required a set of processes that helps to address changes in a systematic way, so it is necessary to consider project management practices (such as agile or predictive project management implementation), a project lifecycle (that allows to make decision and change on time), change management practices (to align people with business perspectives and vision and reduce change barriers), and a commitment from high-level management.

- Enablers [24]: Identify and align with the business vision the needs required to achieve strategic objectives regarding processes, technology, structure, and people. In terms of technology, most of the main enablers used in transport and logistics industry are related to technologies like e-commerce, Intelligent Transportation Systems (ITS), Robotics Process Automation (RPA), Predictive maintenance, drone supervision, Blockchain (DLT), Artificial Intelligence (AI), Analytics, Augmented Reality (AR), and/or Virtual Reality (VR), among others.

- Key Performance Indicators [25]: Known as Key Performance Indicators (KPI), they need to be considered to measure value realization and effectiveness of digitalization. Most of the KPIs used as part of digitalization initiatives for the business case are Net Present Value (NPV), Return on Investment (ROI), Payback Period (PBP), Earned Value (EV), cost indicators (how much is the investment in resources), time indicators (how long it takes), effectiveness (how well we do it at the first time in compliance with the objectives), and efficiency indicators (how well we use sources).

4.4 Enterprise-software-driven process

Software can be defined widely as a piece of a computer programmed used to execute determined activities or for a specific purpose [25]. An enterprise is an organization involving people or entities working together to achieve common goals (Ibid.). Enterprise software can be defined as a set of computer programmes integrated to add value to the business without human intervention in the middle [20].

Among specific trends related to transport and logistics, companies are looking for the implementation on software focused on [1, 20, 26–31]:

- Warehouse Management System [26]: Known as WMS, its purpose is to support logistics operations through management of inventory, distribution centres, and supply capacity from the reception or production to the endpoint of supply.
- Supply Chain Management System [27]: This is an enterprise software known as SCM and includes the management of the overall supply chain, including manufacturing, inventory, sales, points of sales (POS), purchasing, and distribution.
- Transportation Management System [28]: Also known as TMS, it is used with the purpose of optimize freight and gathering information to control fleet conditions and efficiency.
- Customer Relationship Management [29]: This enterprise system involves the management of customer/consumer relationships such as the tracking and management of customer data, customer master data, customer scoring, customer feedback, customer insights, and customer recommendations.
- Enterprise Resource Planning [20]: An ERP system is the core enterprise software for any company, which records any transaction impacting the company, including financial and accounting, production, sales, and marketing, human resources, reporting, and inventory records.
- Toll Collection [1]: Focused on automatically collecting tolls from vehicles moving through roads, highways, or tunnels to save time and costs related to collections.
- Parking Guidance [1]: Solution based on real-time data whose purpose is to inform drivers where they can park their cars and as a result, improve transport smoothing and congestion in a convenient way.
- Public transportation [1]: Used to gather and analyze data to adjust operations to the needs of the citizens to enable greater efficiency and less congested roads.
- Traffic management [1]: Some of the reasons this technology is implemented are related to increasing transportation network efficiency through real-time information, synchronization of traffic lights, and assignment of street space dynamically.
- Data collection [1]: Implemented to analyze movement and traffic using big data to dynamically react to changing situations when unexpected events happen on the road.
- Integration and Operational Software [30]: There are different capabilities of software such as Robotics Process Automation (RPA), Blockchain (DLT), and Artificial Intelligence (AI). RPA can be used to integrate the different systems across the company. It is a based-rule software that can help to automate processes (with multiple activities) to reduce human intervention with low code

automation, depending on the supplier, to support business activities. Another software is Blockchain, which increases end-to-end security and privacy, efficiency through reduced documentation processes, transparency, and reliability. AI is an umbrella of solutions that can reshape operations. Traffic and networks, which can be managed by using assisted intelligence, automation, augmented intelligence, and autonomous intelligence, depending on business needs.

- Maintenance Solutions [31]: The Maintenance Management System (MMS) helps companies with cost reduction, improvement in service quality, positive employee impact and Corporate Social Responsibility (CSR) and environmental issues by managing the predictive maintenance to reduce downtime and use the equipment without breaks, stabilizing delivery times to ensure all of the companies' fleet is available and ready to be scheduled, reacting to problems with equipment, and increasing positive impact of maintenance on environment and waste management. This software usually is connected to the ERP or the SCM system.

4.5 Machine-driven process

Machine-driven processes are those related to warehousing, mobility, rail, and those processes involved in the core operations of a third-party logistics (3PL) company. Some of the benefits of machine-driven core processes are related to the increase of efficiency of deliveries and warehousing, but it is required technology to achieve those objectives. Some of the solutions involved in this trend are [1–3, 32–35]:

- Warehousing robotization [32]: Robotization is the use of physical robots in logistics to handle warehouse materials, including the use of drones. Among applications of robotization in warehousing are product quality check, sorting, intra-warehouse transport, picking, cargo, loading and unloading, and delivery. These solutions help to plan and reduce costs and make decisions, depending on the company objectives.

- Electromobility [33]: Solution used to moderate the impact on transport, warehousing infrastructure, and road transport in the long term to reduce costs. It implies the use of electric-powered vehicles to reduce the environmental hazards resulting from excess fossil fuels burnt that leads to emission restrictions, fines/penalties.

- Warehousing support with AR/VR [34]: Solutions focused on the implementation of devices of augmented reality and mixed reality to improve effectiveness in supply chain management for loading and unloading of materials, order picking, intra-warehouse transporting, security, information handling, information display, complex layout visualization, training, workflow simulation, remote repairs and maintenance, portable control panel, and off-site workstation.

- High-speed rail [1–3]: Solution whose main objective is to reduce time and costs related to the transport of people.

- Last mile delivery optimization [1–3]: These solutions are focused on companies whose main activities are from postal and Courier, Express, and Parcel (CEP) segment) and it is focused on reducing costs while optimized delivery times.

5. Discussion

5.1 Supply chain transformation case study

Through the years, the company had organic growth, as well as acquisitions (inorganic growth), increasing complexity, disparate functional and regional silos, as well as different set of optimized processes. The business challenge was to implement a global ERP system, combining a single global process of demand planning.

Business requirements were complex and required an enterprise-wide supply chain transformation, which was fulfilled by using the following approach:

- Enterprise design focused on the main major business clients and geographies (20%), which represented 80% of common core processes.
- Re-usable project plans and business processes.
- Consolidated product lines into a single set of business processes used as the foundation for business transformation.
- Supply chain processes are implemented in different countries: the United States, Canada, the UK, Ireland, the Netherlands, and Belgium.
- Re-engineering and implementation of demand planning and forecasting processes, as well as, inventory planning, supply planning, production planning, transportation planning, and logistics execution.
- Artificial Intelligence implementation for customer centricity.
- As a result of the ERP implementation and process optimization across overall organization, it was obtained the following benefits:
- Complex enterprise-wide business transformation.
- Increased integration between business units and momentum, enabled by a unified corporate culture.
- Consistent supply chain processes and technology for customers and suppliers.
- Operating model based on IT best practices to excel organization capabilities.
- Customer centricity and consistency increased, service levels increased, and reduction of costs.
- Smart logistics flows allowed a connected supply chain ecosystem.
- Closed loop for integrated planning, digital procurement, and improvement of product development.
- Planning the retirement of over 700 legacy IT systems.

- This case of study was driven by technology, having an indirect impact in climate change because of improvement operations, consolidation, and connectivity among countries, which is one of the big benefits of the current transformation trends from **Figure 2**.
- Global trade.
- Market's domestic.
- Digitalization.
- Enterprise Driven Software.

5.2 Environmental footprint reduction

A sportswear company recognized globally wanted to understand and track its environmental impact through its value chain. The company wanted technology to understand the impact drivers, the consequences of actions related to impact mitigation, and capabilities to cascade targets across business areas. They had implemented a track record for innovative sustainable products, being a recognized leader focused on sustainability; however, they need to improve the sustainable business model to track its performance against past years and targets.

Based on the business requirements, it was designed and built a transparent, fully customizable, and interactive footprint tool, was integrated into the legacy Information Technology (IT) infrastructure with the following characteristics:

- A detailed and comprehensive environmental footprint through a natural capital evaluation approach is used to set a monetary value for the impact on society and the environment so the targets can be measured quantitatively.
- Sourced data on the environmental impact of cradle to grave the impact of key materials, processes, logistics, and consumer behaviors.
- Data adaptation according to the geographic regions of the business.
- The driver for this company was the environmental impact of their operations, with the following benefits:
- Performance track against past years.
- Continuous tracking of their corporate sustainability performance.
- Confidence in using its environmental footprint to set future business targets.
- Continuous measurement tracking of its performance to understand the consequences of its actions.
- This company was focused on a decarbonization strategy to reduce its impact on climate change, but it also transformed and executed processes related to data and technology, which helps to understand how it is working in the digital and

information economy era and how it can be used to improve business performance through the digitalization trend.

5.3 Net zero emissions

Business was looking for an aligned roadmap to its strategy to deliver its public commitment for Net Zero emissions as part of its Vision, ambition and targets, decarbonization strategy, and business transformation. This company has over 150,000 employees over 120 countries, which made a complex transformation process to continue being a leading company in the "post-carbon" era. To achieve this goal, a roadmap for the comprehensive Net Zero strategy was required, along with hands-on transformation support and upskilling the business.

The implemented approach was based on the understanding of the strategic business direction, maturity, and baseline for which it was carried out:

- Maturity assessment with a Net Zero framework will be used to build the new strategy.
- Benchmarking against leading peers based on Net Zero transformation practices.
- Set up tools and methodologies to baseline Scope 1 (Direct emissions owned or controlled by the company), Scope 2 and Scope 3 emissions (Indirect emissions consequence of the company's activities which come from sources not owned or controlled by the organization).
- Implementation of bespoke cloud-based carbon reduction tracker and carbon calculator to report and measure on carbon reduction.
- Staff upskilling to identify, generate and execute on carbon reduction.
- Strategy execution brought benefits not only related to decarbonization but also created different impacts across business, such as:
- Net Zero strategy that led to strengthened market leadership.
- Carbon reduction projects on Scope 1 and Scope 2.
- Baselined Supply Chain Spend across Tier 1 to Tier 3 and developed calculation methodologies and templates of product sales.
- Development of a Task Force on Climate-Related Financial Disclosures (TCFD) across over 900 locations, identifying transition risks and opportunities.
- Embedded information governance architecture, automating Key Performance Indicators (KPIs) and correcting auditing findings.
- Upskilled teams across the business to deliver Net Zero transformation.
- Improvement of Earnings Before Interest and Taxes (EBIT) to fund transformation.

- Transformation of Supply chain and procurement, Operations & Logistics, commercial, people, and culture, data and technology, and Backoffice processes.
- Clear roles and responsibilities for People (Social) and Principle (Governance) related to Net Zero Emissions (Environmental).

In this case study, it can be seen the concurrence of all the trends in **Figure 5** (Global Trade, Market's Domestic Commerce, Digitalization, Machine-Driven Processes, and Software Driven Processes), even though it was strategically focused on decarbonization, helping the company to reduce costs because of the overall impact across the operations.

6. Conclusion

Transport and logistics companies are being impacted in the post-pandemic world, but it is not the only challenge they are facing, since there are other factors impacting their operations, so they need to be more strategic in their investments. Hence, depending on the business's strategic objectives and vision, companies need to identify the best practices aligned with their needs to add value to the final customer through the strategic, financial, operational, legal, regulatory, and compliance areas. There is no one solution, since every option needs to be adapted to the business needs. It means that even the five big trends identified are classified as global trade, market's domestic commerce, digitalization, enterprise-software-driven processes, or machine-driven processes, every company need to assess what is its current state and identify gaps to achieve future state to implement what is best for the business, identifying capabilities related to process, technology, people, and organizational structure to add value to the final customers by leveraging on the current capabilities while developing skills and competencies for the future state and achieve competitive advantage. Moreover, through the cases of study, it can be stated that these trends lead to the value creation for the business, and they are linked to one another, improving efficiency, effectiveness, and productivity, but also, they are focusing on reducing climate change, which is one of the current business imperatives, as for logistics as for any other industry.

Author details

Felix-Eduardo Bueno-Pascual
Information Technologies, Universidad TecMilenio, Mexico City, Mexico

*Address all correspondence to: felixbueno.docente@gmail.com

References

[1] PwC. 25th CEO Survey: Reimagining the Outcomes that Matter [Internet]. 2022. Available from: https://www.pwc.com/mx/es/ceosurvey/2022.html

[2] SYSPRO. A Guide to Understand the Porter's Value Chain and ERP [Internet]. 2018. Available from: https://www.syspro.com/dl/WP/SYSPRO-Understanding-The-Porters-Chain-Guide.pdf

[3] Placek M. Logistics Industry Worldwide – Statistics & Facts [Internet]. 2023. Available from: https://www.statista.com/topics/5691/logistics-industry-worldwide/

[4] Blaga L. Logistics and supply chain management: An overview. Sciendo. 2019;**14**(2):209-215. DOI: 10.2478/sbe-2019-0035

[5] Bueno-Pascual FE, Martinez-Flores JL, Sanchez-Partida D, Cano-Olivos P. Vehicle assignment with penalty costs: A case of Mexican freight transport. Acta Logistica. 2022;**9**(1):75-84. DOI: 10.22306/al.v9i1.272

[6] World Bank Group. Globalization, growth, and poverty: Building an inclusive world economy (English) [Internet]. Available from: https://documents1.worldbank.org/curated/en/954071468778196576

[7] Indian Institute of Materials Management. Logistics and Warehousing Management. India: Indian Institute of Materials Management; 2020. ISBN: 978-93-89795-50-9

[8] PwC. PwC's Global Annual CEO Survey Report [Internet]. 2021. Available from: https://www.pwc.com/gx/en/ceo-agenda/ceosurvey/report-archive.html

[9] PwC. PwC's 27th Annual Global CEO Survey: Thriving in an age of continuous reinvention [Internet]. 2024. Available from: https://www.pwc.com/gx/en/issues/c-suite-insights/ceo-survey.html

[10] United Nations. The Sustainable Development Goals Report [Internet]. 2023. Available from: https://unstats.un.org/sdgs/report/2023/The-Sustainable-Development-Goals-Report-2023.pdf

[11] PwC. Transport & Logistics Barometer: 2022 Mid-Year Analysis [Internet]. 2022. Available from: https://www.pwccn.com/en/transportation/transport-and-logistics-barometer-nov2022.pdf

[12] E&Y. Trucking & Logistics Update [Internet]. 2022. Available from: https://www.ey.com/en_gl/transportation

[13] Deloitte. Global Transportation Trends 2022: Future-Ready Transportation [Internet]. 2023. Available from: https://www2.deloitte.com/content/dam/Deloitte/tw/Documents/public-sector/pr20230213-global-transportation-trends-2022-report-en.pdf

[14] KPMG. Logistics Services Sector Update [Internet]. 2022. Available from: https://corporatefinance.kpmg.us/insights/2022/transportation-logistics-services-sector-update.html

[15] UNCATD. Key Statistics and Trends in International Trade. 2020 [Internet]. Available from: https://unctad.org/system/files/official-document/ditctab2020d4_en.pdf

[16] Beier M, Hauser C, Hogenacker J. Domestic business-to-business

relationships and the internationalization of SMEs: Evidence from Switzerland. International Journal of Entrepreneurial Venturing. 2016;**8**(1):84-101. DOI: 10.1504/IJEV.2016.074402

[17] Andonov A, Dimitrov GP, Totev V. Impact of E-commerce on business performance. TEM Journal. 2021;**10**(4):1558-1564. DOI: 10.18421/TEM104-09

[18] Gomes AC, Francisco Bezerra de Lima RDS Jr, de Souza Oliveira PR, de Oliveira DA, Siqueira RM, da Silva Nora LAR. Logistics Management in Ecommerce: Challenges and opportunities. Revista De Gestão e Secretariado. 2023;**14**(5):7252-2272. DOI: 10.7769/gesec.v14I5.2119

[19] Sulkoswki L, Koilasinska-Morawska K, Brzozowska M, Morawski P, Schroeder T. Last mile logistics innovations in the courier-express-parcel sector due to the COVID-19 pandemic. Sustainability. 2022;**14**(13):8207-8232. DOI: 10.3390/su14138207

[20] Bueno-Pascual FE. Adoption drivers in decision making of Enterprise software systems implemented in Mexican SMEs. Science Journal of Business Management. 2023;**11**(4):164-173. DOI: 10.11648/j.sjbm.20231104.14

[21] Rachinger M, Rauter R, Müller C, Vorraber W, Schirgi E. Digitalization and its influence on business model innovation. Journal of Manufacturing Technology Management. 2018;**30**:1143-1160. DOI: 10.1108/JMTM-01-2018-0020

[22] García Guiliany J, Paz Marcano A, Suarez BH. Elementos de la Filosofía de Gestión que Sustentan la Competitividad en Empresas del Sector Construcción. Revista de Ciencias Sociales (Ve). 2022;**XXVIII**(5):184-197. ISSN: ISSN-E: 2477-9431

[23] Hensellek S. Digital leadership: A framework for successful leadership in the digital age. Journal of Media Management and Entrepreneurship (JMME). 2020;**2**(1):55-69. DOI: 10.4018/JMME.2020010104

[24] Setiawan I, Purba HH. A systematic literature review of key performance indicators (KPIs) implementation. Journal of Industrial Engineering & Management Research. 2020;**1**(3):200-208. DOI: 10.7777/jiemar.v1i2

[25] Haro AF, Martinez JE, Chango TS, Zambrano TP, Zambrano MF. Enterprise resource planning (ERP) processes for optimal and efficient implementation. Prometeo Conocimiento Científico. 2023;**3**(1):1-13. DOI: 10.55204/pcc.v3i1.e21

[26] Andiyappillai N. Factors influencing the successful implementation of the warehouse management system (WMS). International Journal of Applied Information Systems. 2020;**12**(35):30-24. ISSN: 2249-0868

[27] Basuki M. Supply chain management: A review. Journal of Industrial Engineering and Halal Industries. 2021;**2**(1):9-12. E-SSN: 2722-8142

[28] Sattayathamrongthian M, Vanpetch Y. Business's transportation management system technology adoption in Nakhon Pathom Thailand. Transport Research Procedia. 2022;**63**:2449-2457. DOI: 10.1016/j.trpro.2022.06.281

[29] Idzikowski A, Kurylo P, Cyganiuk J, Ryczko M. Customer relationship management (CRM) philosophy and its significance for the enterprise. Sciendo. 2019;**1**(1):1004-1011. DOI: 10.2478/czoto-2019-012

[30] Silva Acosta DA, Mamede HS, Silva MM. Robotic process automation (RPA) adoption: A systematic literature review. Engineering Management in Production and Services. 2022;**14**(2):1-12. DOI: 10.2478/emj-2022-0012

[31] Shaheen BW, Németh I. Integration of maintenance management system functions with industry 4.0 technologies and features – A review. PRO. 2022;**10**(2173):1-29. DOI: 10.3390/pr10112173

[32] Dhaliwal A. Rise of automation and robotics in warehouse management. In: Transforming Management Using Artificial Intelligence Techniques. India: Manav Rachna International Institute of Research and Studies; 2020. DOI: 10.1201/9781003032410-5

[33] Chinoracky R, Stalmasekova N, Corejova T. Trends in the field of electromobility – From the perspective of market characteristics and value-added services: Literature review. Energies. 2022;**25**(6144):2-19. DOI: 10.3390/en15176144

[34] Husár J, Knapcikova L. Possibilities of using augmented reality in warehouse management: A study. International Scientific Journal About Logistics. 2021;**8**(2):133-139. DOI: 10.22306/al.v8i2.212

[35] Morillo-Aguilar KR. Impact of technological enablers in the industry and manufacturing of SMEs. Ciencia América. 2022;**11**(2):1-12. DOI: 10.33210/ca.v11i2.394

Chapter 4

Advanced Sustainable Logistics with HSR for the Development in Great Montreal Area

Yonglin Ren, Xinyue Ren and Anjali Awasthi

Abstract

Currently, the new transportation tool High-Speed Railway (HSR) pushes economic and social development to a great level in some countries. Because of its high speed (actual speed 430 km/h, experiment speed 600–1200 km/h) and high efficiency, it makes good transportation in a surprisingly quick increment and then supports supply chain logistics running at a greatly higher level than that before. Especially, a stimulation of trade volume will happen due to the increased speed of transportation within the HSR network. The success of HSR in the Asia area implies its future application may produce an economic engine in East Canada or the Great Montreal Area with extended regions, which will stimulate the local economic and social development in an excellent model. Especially for the goods flow or trade volume, the implementation of the HSR network centred in the Great Montreal Area can bring to the community. This chapter will make a mathematical model deviated from the Gravity Model to investigate the relationship between the goods flow and the HSR speed. The research on their relationship demonstrated that the HSR would be able to substitute the low-speed vehicle style and increase economic development.

Keywords: HSR, sustainable, logistics, gravitational model, great Montreal area

1. Introduction

It can list a series of problems coming from current city logistics, which are those conflicts between poor transportation and sustainable city logistics that people pursue. Low-speed transportation always happens because heavy circulation pushes into the same narrow road in a short time for circulation, which produces problems such as the gas assumption that pollutes the air and noise damaging the environment, as a result of low efficiency in GDP development for the reason of low speed of goods and trade among cities. Sustainable advanced city logistics with HSR could solve some city circulation problems of the above. In the past, some problems existed in the limited science and techniques that could not be applied to sustainable city logistics. Currently, sustainable advanced city logistics not only improves facilities and fine policies of the authority but also is impelled by new HSR tools.

This chapter will apply a mathematical model to the simulation of the city trade logistics with HSR to look for its logical-mathematical model and apply for AI to simulate the city logistics that state the application with the HSR relative to the trade, which points out that the HSR improves the city logistics by its high speed with high efficiency for business activities. Traditionally, the plane is the quickest travel tool. However, an electrical high-speed railway (EHSR) is quicker than a plane in a short distance if the travel time is limited to 4 hours then the plane needs at least 2 hours for its boarding process, though the speed of HSR is 420 k/h and the plane is 700 k/h. This chapter will make a mathematical model to evaluate the influence and result of the application of the HSR for the trades in the Great Montreal Area (GMA). HSR can provide sustainable city logistics not only in the city area but anywhere that can be defined as city logistics. This chapter investigates what and how an HSR with smart algorithms improves sustainable city logistics in the field of trade circulation. By the data analysis, the mathematical model of the HSR in sustainable city logistics could predict a good advantage for the GDP increase in the GMA. And according to the load and the speed, other transportation tools cannot compete with HSR in terms of efficiency [1]. Especially, the higher the speed of HSR developed, the more advantages it has, and no other transportation will be comparable to such HSR [2].

When we are talking about interprovincial and international trade, speed is a key to an efficient supply, which leads to an increased demand and thus, an improvement of the local economy. This chapter will discuss first the implementation of the factor speed to the Gravitational Model for trade, then an analysis of the Great Montreal Area, including different factors that can be included in the general Gravitational Model to make it suitable for the Great Montreal Area. This chapter will compare the before and after of the implementation of HSR in the relationship with trade.

2. Literature

Sustainable advanced city logistics in the shortest time has been researched by many researchers. How to spend less time making the best city logistics service? Meituan company applies smart techniques in its delivery service in an average of 28 minutes from the provider to the customer, in which AI is used for the service among the components including the managers, delivers, providers, and customers, which benefits all shareholders among the logistics service system. However, sustainable city logistics in the suburbs are not like the urban area, which is a desolate area to make the service more difficult. To solve the problem, in 1959 in Japan, it constructed a high-speed railway (HSR) from Tokyo to Osaka, which made the city logistics better in suburban areas than before. For example, now it only takes 5 hours from Shanghai to Beijing to pass 1213.0 km. China High-speed Railway (CHSR) not only makes city logistics better at a great level in remote areas, but it also provides a sustainable solution for development in remote village areas for trades and travelers.

The sustainable advanced city logistics applied for an algorithm of the metaheuristics by Gogna & Tayal [3] is based on AI technology and neural networks algorithms. Holland [4] also proposed a genetic algorithm (GA) based on principles from evolution theory, which defined genetic algorithms originated from the Monte Carlo method. Murthy & Chowdhury [5] integer encoding of chromosomes used in the supply chain logistics then it developed this simulated algorithm (SA) approach for examining the equations of state and frozen states of n-body systems for the optimization of routing net. One algorithm appears to be available stochastic algorithms for global optimization from Ali & Storey [6] which could support better choice of advanced city logistics. And the advanced city logistics could applied for Ant Colony Optimization (ACO) by Marco [7] and Anirudh [8], which stated that ACO is a positive feedback and is efficient for traveling salesman problems (TSP) in the advanced city logistics. The author Zhang et al. [9] improved the logistics strategy and proposed a new Route Decomposition (RD) and a Memetic Algorithm (MA) framework for the Periodic Capacitated Arc Routing Problem (PCARP).

The advanced city logistics in cities mainly focus on the flow of trade and human beings, which is so complicated that many problems result from the heavy circulation thus causing congestion in many metropolitan areas, especially in the rush hours. Kostas [10] researched the key challenges associated with adopting, designing and managing performance-based contracts (PBC) for advanced city logistics services, in which mostly the PBC originates in upstream supply chain resources rather than downstream initiatives. The researchers Meng & Yari [11] studied the advanced city logistics for intelligent and sustainable transport by investigating the global developments in transport and city logistics and the cost-efficiency and flexibility of European transport and city logistics to show the customers' requirements in various fields, in which it produces challenge from the relationship of sustainability issues of social, economic, and environmental parameters. Those researchers Takai Eizo [12] studied advanced city logistics through the role of operations, in which the problems in city logistics in Japan are discussed and a solution for the problem by assumptions of some cause-and-effect relationship alone is not sufficient. Advanced city logistics to improve and optimize delivery management in a smart environment process by Yassine et al. [13] was studied in which the manufacturing of products and services are sold online in various regions and transporters are principally demanded to apply the optimum types of shipping centres and shipping routines, and smart delivery city logistics should be able to meet those aspects: the delivery requirements. Masato [14] attempts to contribute to transcending the stereotypes of value systems in shipping by presenting some of the elements of current maritime. In the dynamic infrastructure, Miyashita Kunio [15] studied the structural change in international advanced city logistics, in which the type of contractual system utilizes mid to long-term contracts to shipper satisfaction because of the guarantee of long-term profit stability. Under specific conditions, selected cargo flows show the potential to be shifted to other modes of transport or at least other types of vehicles that are more suitable for operations in a dense urban environment, in which voluntary cooperation of freight operators may be considered a method of improving the effectiveness of city logistics operations from Kaszubowski [16] and different regions may take various measures and city logistics and delivery services have been optimized, in which the costs need to be optimized elsewhere. Those researches relate an important factor to the trade and human beings flow, in which the advanced logistics needs the help from HSR to solve its efficiency and help the economic development.

3. The gravitational model

The gravitational model predicts the value of trade between places i and j in relationship with these places' GDP and distance between these two places [17]:

$$T_{ij} = \frac{AY_i^a Y_j^b}{D_{ij}^c} \quad (1)$$

where T_{ij} is the value of trade between places i and j.
A, a, b, c are constants.
Y_i is the GDP of place i.
Y_j is the GDP of place j.
D_{ij} is the distance between i and j.

In general, D_{ij} can be calculated by considering the actual distance, the cultural affinity, geographic differences, presence of multinational corporations and borders between places i and j. This chapter probes the possibility of replacing D_{ij} with the travel time between places i and j (t_{ij}) with a change of A.

$$T_{ij} = \frac{AY_i^a Y_j^b}{D_{ij}^c} = \frac{BY_i^a Y_j^b}{t_{ij}^c} \quad (2)$$

where $A/D_{ij}^c = (A/s_{ij}^c)(1/t_{ij}^c) = B/t_{ij}^c$;
S_{ij} is the speed of the transportation vehicle between places i and j.

The new equation is mathematically logical as speed = distance/time, thus distance = speed*time.

The new equation also makes sense, for Y_i, Y_j, D_{ij} large, as the more time it takes to travel from places i to j, the more reluctant people tend to trade from places i to j as the relative monetary and time cost increase. With the same logic, in contrast, the less time it takes to travel from places i to j, the value of trade increases as trade demand and supply are more efficient.

4. Montreal trade data

In order to construct a Gravity Model suitable for the Great Montreal Area—finding B, a, b and c—we need to consider several data in order to achieve the goal using simulation and AI analysis.

As the most recent data available for our purpose is in 2014, all data will be in 2014. Note that as the Great Montreal Area does not have HSR yet, the choice of year does not matter as long as it is prior the date of this chapter's publication, as the HSR influence over data does not exist. However, more recent data can lead to more precise final results in terms of simulated trade volume as they are the results of other factors that improve the trade.

First, the volume of trade between the Great Montreal Area with its trade partners is main trade in Quebec. Note that as the public information contains only the information about the Great Montreal Area's province, Quebec, consider Quebec's first. **Table 1** contains the data of Quebec trade partners—10 Canadian provinces, 3 Canadian territories and top 10 US state trade partners. As the above trade data is not available for the Great Montreal Area, we can estimate instead the Great Montreal Area's in proportion of its GDP ratio over Quebec's.

$$\text{Trade}_{\text{Great Montreal Area,Quebec's Trade Partner}}/\text{Trade}_{\text{Quebec,Qubec's Trade Partner}} = \text{GDP}_{\text{Great Montreal Area}}/\text{GDP}_{\text{Quebec}}$$

DOI: http://dx.doi.org/10.5772/intechopen.1005318

Thus, $\text{Trade}_{\text{Great Montreal Area, Quebec's Trade Partner}}$

$$= (\text{Trade}_{\text{Quebec,Qubec's Trade Partner}} * \text{GDP}_{\text{Great Montreal Area}} / \text{GDP}_{\text{Quebec}}$$
$$= \text{Trade}_{\text{Quebec,Qubec's Trade Partner}} * (190{,}227{,}000{,}000\ \$\text{CAD}\ [23]/338319000000\$\text{CAD}\ [24])$$
$$= \text{Trade}_{\text{Quebec,Qubec's Trade Partner}} * 0.56227111$$

With the estimated ratio (**Table 2**)

Now, consider the average time it takes for trade to arrive from the starting point to the destination [23, 24].

The most recent data about the repartition of transportation modes in Canadian trade is 2011s (**Table 3**).

Trade Partners	Export from Trade Partners (in millions of $CAD)	Import from Trade Partners (in millions of $CAD)	Total Trade with Trade Partners(in millions of $CAD)	GDP of Trade Partners(in millions of $CAD) [19–21][1]
Newfoundland and Labrador	1897	2014	3911	32136.7
Prince Edward Island	380	260	640	5332
Nova Scotia	2581	1511	4092	36258.3
New Brunswick	3254	3565	6819	29,650
Ontario	38,644	42,087	80,731	676838.4
Manitoba	2172	2030	4202	59499.7
Saskatchewan	1942	944	2886	78506.4
Alberta	9737	6350	16,087	365191.1
British Columbia	6215	4157	10,372	225863.3
Yukon	71	18	89	2650.3
Northwest Territories	221	87	308	4656.5
Nunavut	297	38	335	2350.9
Texas	3575.147956	7709.004393	11284.15235	1781770.194
New York	6036.504659	2038.388785	8074.893444	1535564.161
Vermont	3412.103269	1383.52492	4795.628189	32409.32894
Pennsylvania	3362.580985	1215.489027	4578.070012	755902.3393
Ohio	3449.320952	754.691009	4204.011961	652916.6188
New Jersey	2304.565172	1086.196318	3390.76149	598898.127
Connecticut	2467.022152	615.315773	3082.337925	270207.2607
Massachusetts	1708.700031	1242.214485	2950.914516	506801.9852
Tennessee	2525.264476	289.61916	2814.883636	329905.1122
California	1566.896317	1190.712692	2757.609009	2593340.192

[1]*GDP: For Canadian provinces: https://www150.statcan.gc.ca/t1/tbl1/en/tv.action?pid=3610048701; For US provinces: https://hub.arcgis.com/datasets/a2a8826c4ba44f89a684000fe7c05b8c_0?geometry=-126.256%2C20.471%2C1.185%2C63.314&selectedAttribute=GDP_by_State___97_to_16__dol_39; adjusted for 1 $CAD = 0.905912 $US in 2014 for exchange based on https://www.ofx.com/en-ca/forex-news/historical-exchange-rates/yearly-average-rates/*

Table 1.
Data of Quebec's trade partners in 2014 [18].

Trade Partners	Estimation of Total Trade with Trade Partners (in millions of $CAD)	GDP of Trade Partners (in millions of $CAD)	Distance from the Great Montreal Area to the Trade Partners (in km) [22]
Newfoundland and Labrador	2199.042315	32136.7	1426
Prince Edward Island	359.853511	5332	1136.1
Nova Scotia	2300.813386	36258.3	1242.9
New Brunswick	3834.126706	29,650	748.7
Ontario	45392.70906	676838.4	1076
Manitoba	2362.663208	59499.7	2014.75
Saskatchewan	1622.714426	78506.4	3168.9
Alberta	9045.255363	365191.1	3838.6
British Columbia	5831.875963	225863.3	3899
Yukon	50.04212888	2650.3	4222
Northwest Territories	173.1795022	4656.5	3744
Nunavut	188.3608222	2350.9	2805
Texas	6344.752878	1781770.194	3083.6
New York	4540.279308	1535564.161	599.6
Vermont	2696.44319	32409.32894	172.6
Pennsylvania	2574.116512	755902.3393	701.6
Ohio	2363.794476	652916.6188	1108.5
New Jersey	1906.52723	598898.127	703.4
Connecticut	1733.10957	270207.2607	517.9
Massachusetts	1659.213983	506801.9852	503.5
Tennessee	1582.727749	329905.1122	1863.1
California	1550.523881	2593340.192	4790.8

Table 2.
Data of great Montreal Area's trade partners in 2014.

Transportation Modes	Repartition (%) [25]	Average Speed (km/h)
Trucking	57.0558956	96.5605 [26]
Air	20.7237506	925.373 [27]
Rail	18.8841057	35.4056 [28]
Marine	3.3362481	4.97097 [29]

Table 3.
Repartition of transportation modes in Canadian trade in 2011.

Thus, we can obtain an average speed for trade transportation:

$$57.0558956\% * 96.5605 + 20.7237506\% * 925.373 + 18.8841057\% * 35.4056 + 3.3362481 * 497097 = 2537173254 \text{km/h}$$

Trade Partners	Estimation of Total Trade with Trade Partners(in millions of $CAD): T_{ij}	GDP of Trade Partners(in millions of $CAD): Y_j	Time taken from the Great Montreal Area to the Trade Partners with an average speed of 253.7173254 km/h(in h): t_{ij}
Newfoundland and Labrador	2199.042315	32136.7	5.620428
Prince Edward Island	359.853511	5332	4.477818
Nova Scotia	2300.813386	36258.3	4.898759
New Brunswick	3834.126706	29,650	2.950922
Ontario	45392.70906	676838.4	4.24094
Manitoba	2362.663208	59499.7	7.940924
Saskatchewan	1622.714426	78506.4	12.48988
Alberta	9045.255363	365191.1	15.12944
British Columbia	5831.875963	225863.3	15.3675
Yukon	50.04212888	2650.3	16.64057
Northwest Territories	173.1795022	4656.5	14.75658
Nunavut	188.3608222	2350.9	11.05561
Texas	6344.752878	1781770.194	12.15368
New York	4540.279308	1535564.161	2.36326
Vermont	2696.44319	32409.32894	0.680285
Pennsylvania	2574.116512	755902.3393	2.765282
Ohio	2363.794476	652916.6188	4.369035
New Jersey	1906.52723	598898.127	2.772377
Connecticut	1733.10957	270207.2607	2.041248
Massachusetts	1659.213983	506801.9852	1.984492
Tennessee	1582.727749	329905.1122	7.343212
California	1550.523881	2593340.192	18.88243

Table 4.
Data for simulation.

Remark that this speed is less than the current average HSR speed—325 km/h [30], much less than 660 km/h and 1000–1400 km/h HSR may exist in the future (**Table 4**).

5. Simulation and AI analysis

Now, using the data, we will run simulations in order to find the constant terms B, a, b and c in the formula

$$T_{ij} = \frac{BY_i^a Y_j^b}{t_{ij}^c}. \tag{3}$$

Yi = Great Montreal GDP = 190,227 M \$CA

The attempt to find constants for formula (3) consists of linear regression [31]. Thus, first, we make the predicted formula (3) linear.

$$T_{ij} = \frac{BY_i^a Y_j^b}{t_{ij}^c} \tag{4}$$

$$\ln T_{ij} = \ln \mathrm{B} + a * \ln \mathrm{Y_i} + b * \ln \mathrm{Y_j} - c * \ln \mathrm{t_{ij}} \tag{5}$$

As we use (5) to estimate the gravity function for the Great Montreal Area i in relationship to another area j, we can set

$ln\ T_{ij} = pred_j$

$ln\ B = a_0$

$a * ln\ Y_i = a_1 * x_1 = a_1 * ln\ (190227)$

$b * \ln \mathrm{Y_j} = a_2 * x_{2j}$

$c * \ln \mathrm{t_{ij}} = a_3 * x_{3j}$

Thus, (5) becomes $\mathrm{pred_j} = a_0 + a_1 * x_{1j} + a_2 * x_{2j} - a_3 * x_{3j}$

To obtain the best-fit constants for the equation using regression, we need to consider the cost function before simulation. The cost function consists of mean square error (MSE) —an average difference square between the estimated total trade volumes and the true ones. As we have 22 trading partners to consider, our cost function is

$$C = \frac{1}{22}\sum_{j=1}^{22}\left(pred_j - y_j\right)^2 \tag{6}$$

Where y_j is the actual total trade with trade partners (in millions of \$CAD) in the table for Data for simulation.

Thus, our goal is to minimize the cost function using simulation

$$\min C = \min \frac{1}{22}\sum_{j=1}^{22}\left(a_0 + a_1 * x_1 + a_2 * x_{2j} - a_3 * x_{3j} - y_j\right)^2 \tag{7}$$

Based on directional derivatives, $\nabla C = \begin{bmatrix} \frac{\partial C}{\partial a_0} \\ \frac{\partial C}{\partial a_1} \\ \frac{\partial C}{\partial a_2} \\ \frac{\partial C}{\partial a_3} \end{bmatrix}$ always points in the direction of the steepest ascent [32]. Hence, the negative∇C will lead to the most descent. Thus, it suffices to use derivatives to update the constants a_0, a_1, a_2 *and* a_3 through gradient descent.

		Number of Simulations (m)	
		10^3	**10^4**
α	10^{-4}	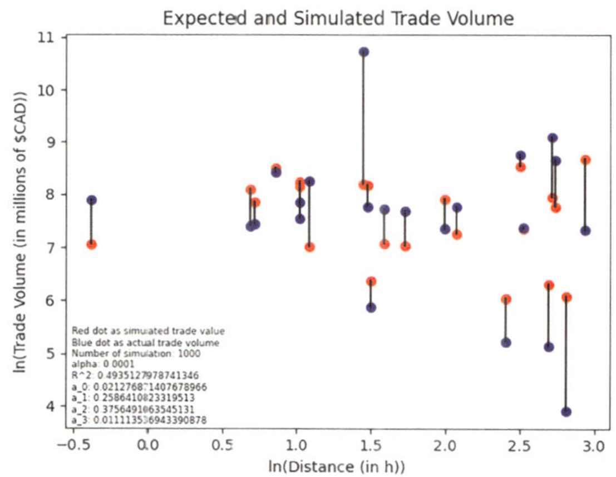 	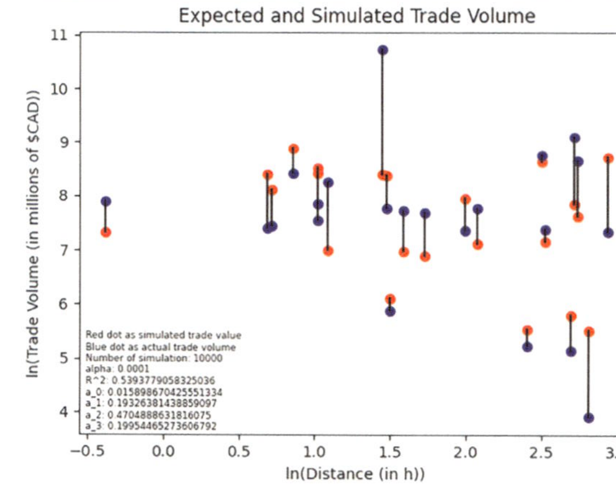
	10^{-5}	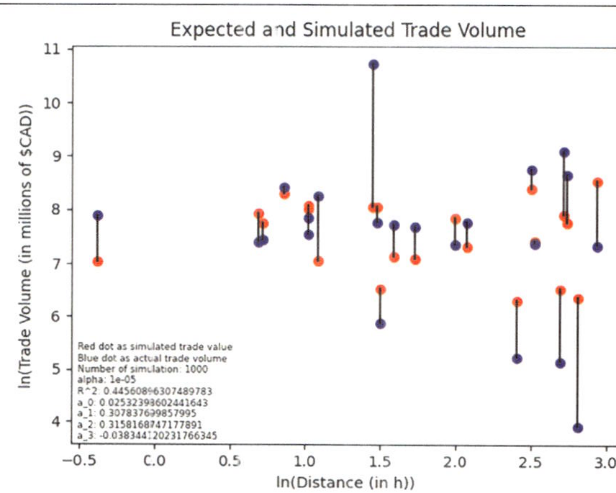 	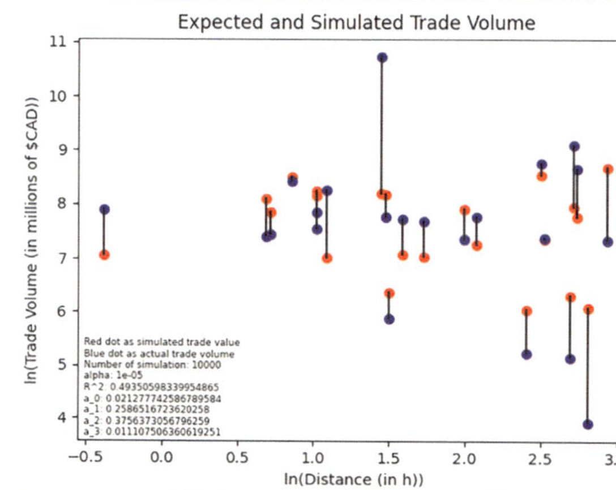

<table>
<tr><td></td><td></td><td colspan="2">Number of Simulations (m)</td></tr>
<tr><td></td><td></td><td>10^3</td><td>10^4</td></tr>
<tr><td>α</td><td>10^{-6}</td><td>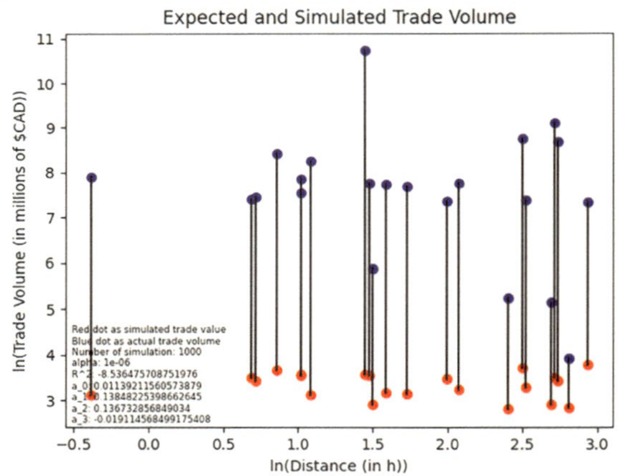
</td><td>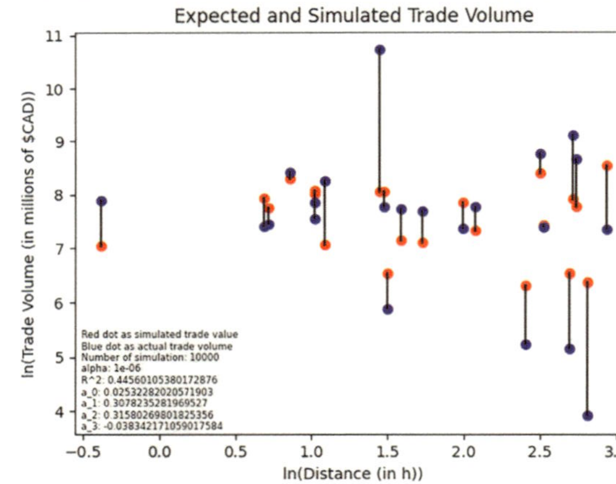
</td></tr>
<tr><td></td><td>10^{-7}</td><td>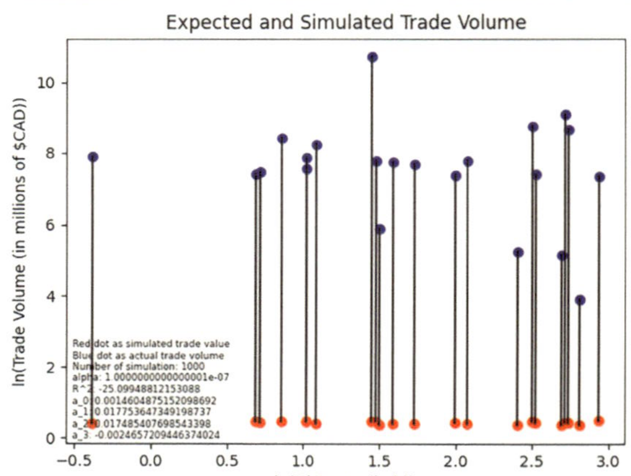
</td><td>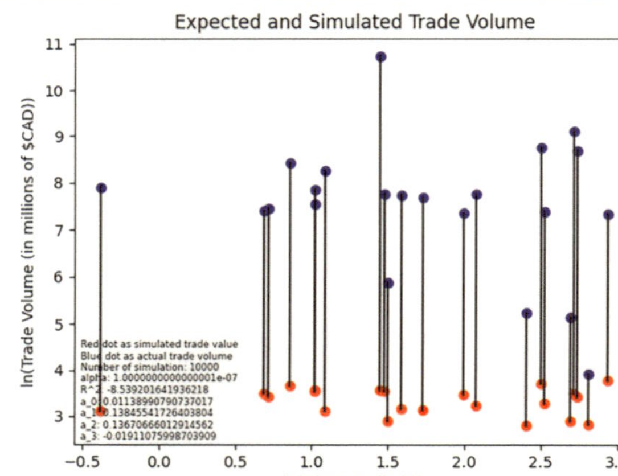
</td></tr>
</table>

		Number of Simulations (m)	
		10^3	10^4
α	10^{-4}	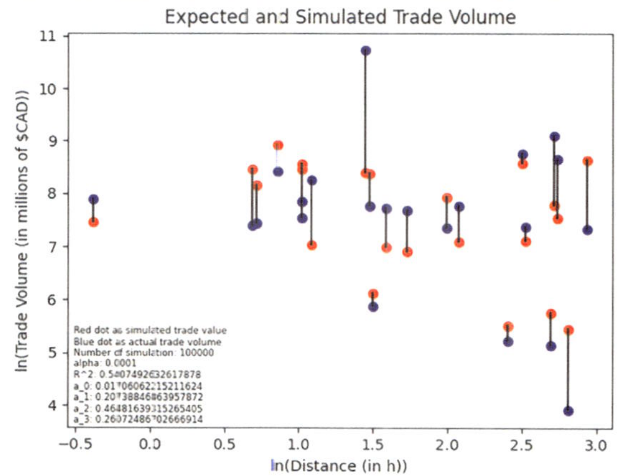	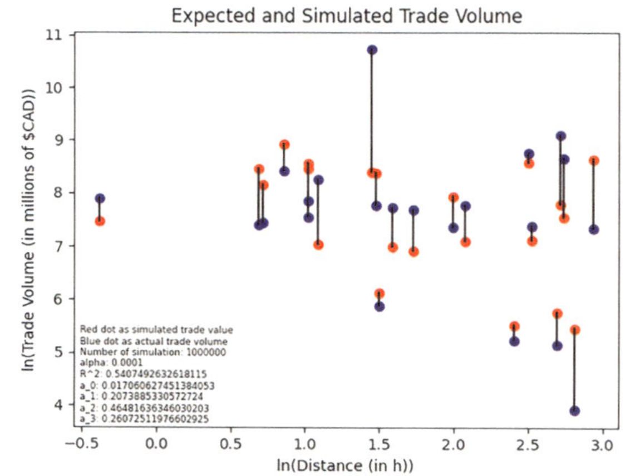
	10^{-5}	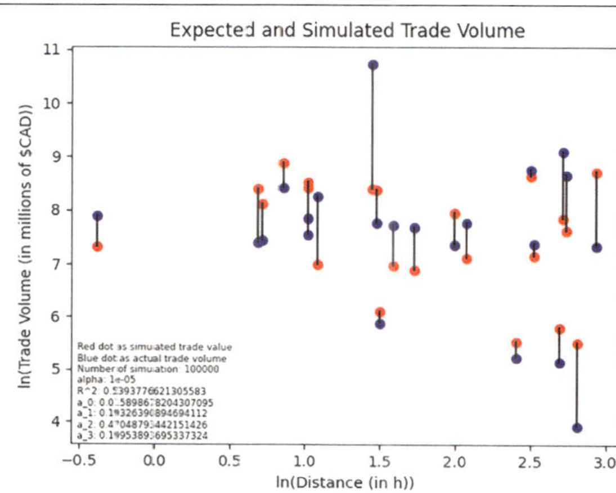	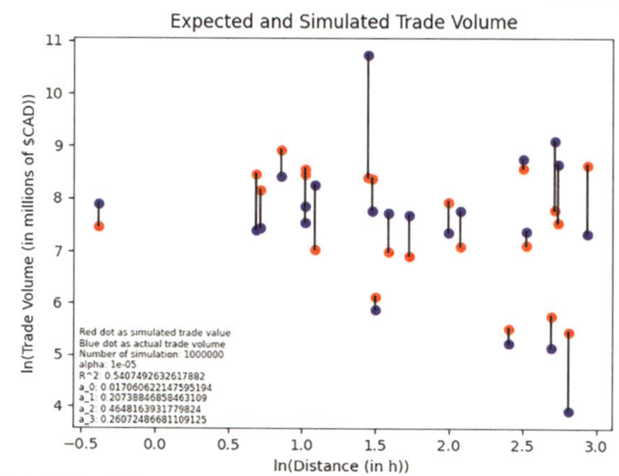

<table>
<tr><td></td><td></td><td colspan="2">Number of Simulations (m)</td></tr>
<tr><td></td><td></td><td>10^3</td><td>10^4</td></tr>
<tr><td>α</td><td>10^{-6}</td><td>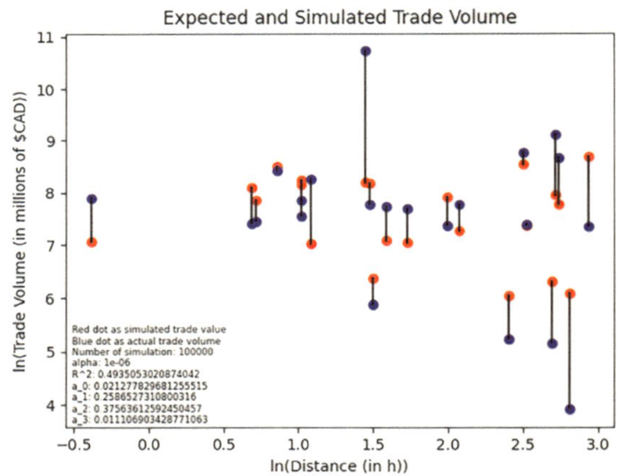
</td><td>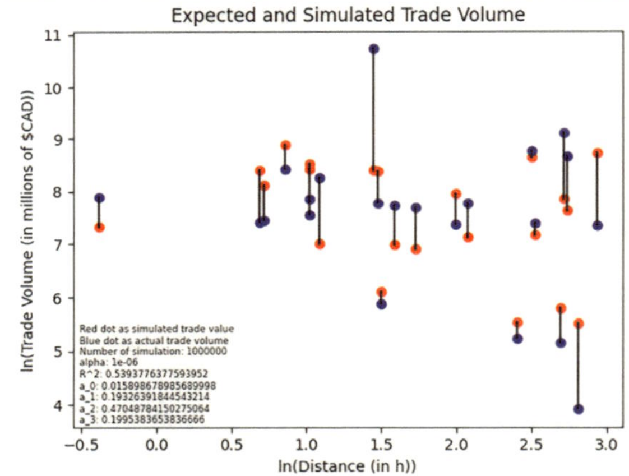
</td></tr>
<tr><td></td><td>10^{-7}</td><td>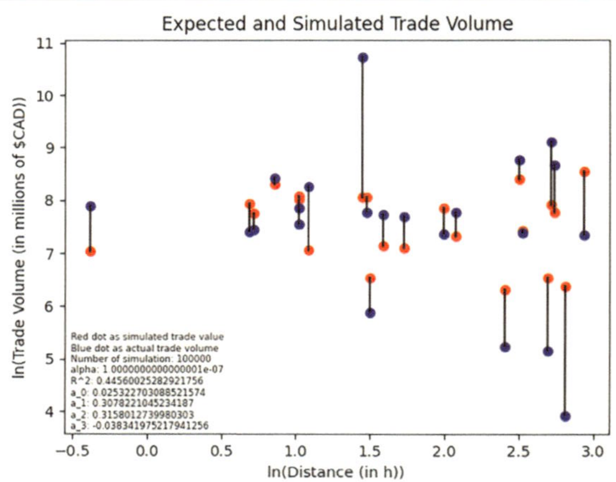
</td><td>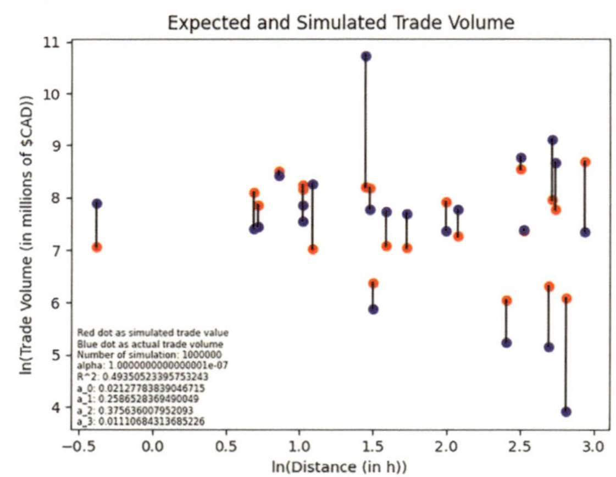
</td></tr>
</table>

Table 5.
Simulation results

DOI: http://dx.doi.org/10.5772/intechopen.1005318

Thus, for $n = 22; learning\ rate\ \alpha\ small$

$$\frac{\partial C}{\partial a_0} = \frac{2}{n}\sum_{j=1}^{n}\left(pred_j - y_j\right); a_0 = a_0 - \alpha\frac{2}{n}\sum_{j=1}^{n}\left(pred_j - y_j\right) \tag{8}$$

$$\frac{\partial C}{\partial a_1} = \frac{2}{n}\sum_{j=1}^{n}\left(pred_j - y_j\right)x_1; a_1 = a_1 - \alpha\frac{2}{n}\sum_{j=1}^{n}\left(pred_j - y_j\right)x_1 \tag{9}$$

$$\frac{\partial C}{\partial a_2} = \frac{2}{n}\sum_{j=1}^{n}\left(pred_j - y_j\right)x_{2j}; a_2 = a_2 - \alpha\frac{2}{n}\sum_{j=1}^{n}\left(pred_j - y_j\right)x_{2j} \tag{10}$$

$$\frac{\partial C}{\partial a_3} = \frac{2}{n}\sum_{j=1}^{n}\left(pred_j - y_j\right)\left(-x_{3j}\right); a_3 = a_3 + \alpha\frac{2}{n}\sum_{j=1}^{n}\left(pred_j - y_j\right)x_{3j} \tag{11}$$

Through simulations, we can update $a_0, a_1, a_2\ and\ a_3$ m times to approximate the min C.

Using the AI in python to simulate,

Trying with several numbers of simulation and α, we obtain (**Table 5**)

By intuition, the gradient leads to the min C with larger m steps with the smallest learning rate

α, which leads the movement and the direction for each step. However, considering the tables, we can clearly see that it is not the case. For example, the result of 10^3 simulations with $\alpha = 10^{-7}$ is worse than smaller m and bigger α. Thus, we can conclude that C is not convex. Consider the best-predicted $a_0, a_1, a_2\ and\ a_3$ with the biggest R^2, the one with 10^6 simulations and $\alpha = 0.0001$ (**Figure 1**).

By Z-score $\frac{x-\mu}{\sigma}$, all differences obtained from ln(trade volume) – ln(simulated trade volume) are within 2σ except the one at ln(distance) = 1.44478 (**Table 6**).

Thus, by excluding the outlier, we obtain (**Figure 2**)

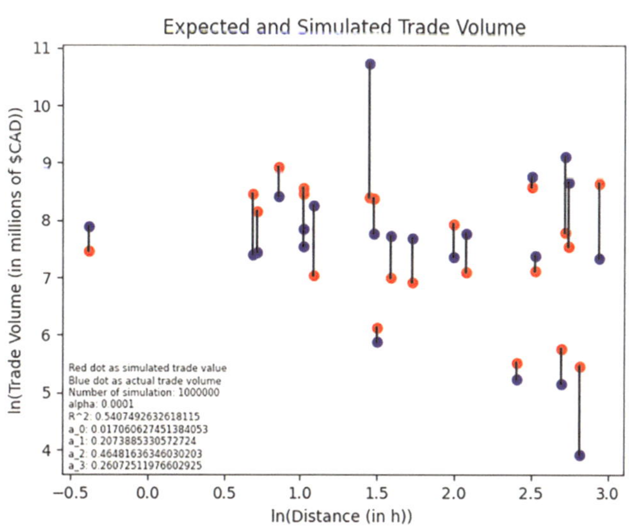

Figure 1.
Simulation result with m = 106 and α = 0.0001.

Standard deviation σ = 0.958541006					
Difference X	**Z score**	**Difference X**	**Z score**	**Difference X**	**Z score**
0.784075	0.817988	1.115281	1.16352	−0.60912	−0.63546
−0.25032	−0.26115	−1.55597	−1.62327	−0.90255	−0.94159
0.737396	0.76929	−0.6078	−0.63409	−0.70779	−0.7384
1.209447	1.261758	−0.28137	−0.29354	−1.05105	−1.09651
2.321481	2.421891	0.178352	0.186066	−0.55755	−0.58167
0.659641	0.688172	−0.51413	−0.53636	−1.29027	−1.34608
0.27318	0.284996	0.433499	0.452249		
1.326773	1.384158	−0.71121	−0.74198		

Table 6.
Difference of non-simulated and simulated trade volume.

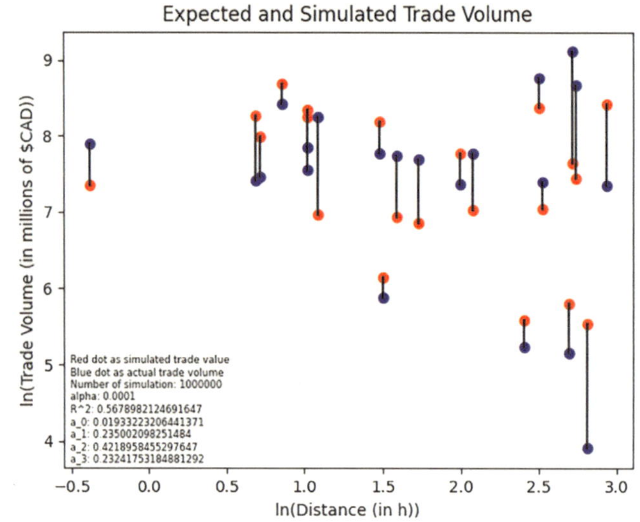

Figure 2.
Simulation result with m = 106 and α = 0.0001 without outlier.

Thus, consider the estimated a_0, a_1, a_2 *and* a_3 and previous settings for $T_{ij} = \frac{BY_i^a Y_j^b}{t_{ij}^c}$, In addition, to consider the influence of noise to the predicted value, we can also add a residual in the formula to the estimated value.

Thus,

$$T_{ij} = \exp\left(ln\ B + a * ln\ Y_i + b * ln\ Y_j - c * ln\ t_i\right) + \in; \tag{12}$$

where $\in$ *=Actual Total trade – Simulated trade value*

$$\mathrm{B} = e^{a_0} = \mathrm{e}^{0.01933223206441371}$$

$$a = a_1 = 0.235002098251484$$

$$b = a_2 = 0.4218958455297647$$

$$c = a_3 = 0.23241753184881292$$

R^2 = 0.5678982124691647 (**Tables 7–9**)

Trade Partners	ln(Total Trade with Trade Partners)(in ln(millions) of $CAD)	ln(Simulated Total Trade with Trade Partners)(in ln (millions) of $CAD)	(Estimated) Total Trade with Trade Partners(in millions of $CAD)	Simulated Total Trade with Trade Partners(in millions of $CAD)	Adjustment ∈(in millions of $CAD)
Newfoundland and Labrador	7.695777233	6.853095312	2199.042315	946.8070388	1252.235276
Prince Edward Island	5.885697035	6.148077493	359.853511	467.8171399	−107.9636289
Nova Scotia	7.741017985	6.935945535	2300.813386	1028.591362	1272.222024
New Brunswick	8.251696971	6.96886162	3834.126706	1063.01195	2771.114756
Ontario	10.72310678	8.204249089	45392.70906	3656.453945	41736.25512
Manitoba	7.76754474	7.032643024	2362.663208	1133.021259	1229.641949
Saskatchewan	7.391855598	7.044336861	1622.714426	1146.348395	476.3660306
Alberta	9.10999563	7.648332464	9045.255363	2097.145607	6948.109756
British Columbia	8.671094005	7.441986786	5831.875963	1706.136584	4125.739379
Yukon	3.912865228	5.548053438	50.04212888	256.7373141	−206.6951852
Northwest Territories	5.154328642	5.813756233	173.1795022	334.8746333	−161.6951311
Nunavut	5.23835939	5.592515625	188.3608222	268.4099901	−80.0491679
Texas	8.755383432	8.367914608	6344.752878	4306.645648	2038.10723
New York	8.420743811	8.685779367	4540.279308	5918.151114	−1377.871806
Vermont	7.899688847	7.347444238	2696.44319	1552.224338	1144.218852
Pennsylvania	7.853261652	8.350251559	2574.116512	4231.245018	−1657.128506
Ohio	7.768023436	8.182151848	2363.794476	3576.542562	−1212.748086
New Jersey	7.553038662	8.251429965	1906.52723	3833.103108	−1926.575878
Connecticut	7.457672513	7.986795255	1733.10957	2941.853943	−1208.744373
Massachusetts	7.414099265	8.258692476	1659.213983	3861.042395	−2201.828412
Tennessee	7.366905061	7.773467941	1582.727749	2376.699266	−793.9715167
California	7.346348141	8.423865627	1550.523881	4554.475371	−3003.95149
Total	165.3785041	162.8596464	100312.1252	51257.338	49054.78719

Table 7.
Data adjustment with residual ∈.

Trade Partners	GDP of Trade Partners(in millions of $CAD): Y_j	Distance from the Great Montreal Area to the Trade Partners (in km)	Time taken from the Great Montreal Area to the Trade Partners with HSR speed of (in h): t_{ij}			
			325 km/h	660 km/h	1000 km/h	1400 km/h
Newfoundland and Labrador	32136.7	1426	4.387692308	2.160606061	1.426	1.018571429
Prince Edward Island	5332	1136.1	3.495692308	1.721363636	1.1361	0.8115
Nova Scotia	36258.3	1242.9	3.824307692	1.883181818	1.2429	0.887785714
New Brunswick	29,650	748.7	2.303692308	1.134393939	0.7487	0.534785714

Trade Partners	GDP of Trade Partners(in millions of \$CAD): Y_j	Distance from the Great Montreal Area to the Trade Partners (in km)	Time taken from the Great Montreal Area to the Trade Partners with HSR speed of (in h): t_{ij}			
			325 km/h	660 km/h	1000 km/h	1400 km/h
Ontario	676838.4	1076	3.310769231	1.63030303	1.076	0.768571429
Manitoba	59499.7	2014.75	6.199230769	3.052651515	2.01475	1.439107143
Saskatchewan	78506.4	3168.9	9.750461538	4.801363636	3.1689	2.2635
Alberta	365191.1	3838.6	11.81107692	5.816060606	3.8386	2.741857143
British Columbia	225863.3	3899	11.99692308	5.907575758	3.899	2.785
Yukon	2650.3	4222	12.99076923	6.396969697	4.222	3.015714286
Northwest Territories	4656.5	3744	11.52	5.672727273	3.744	2.674285714
Nunavut	2350.9	2805	8.630769231	4.25	2.805	2.003571429
Texas	1781770.194	3083.6	9.488	4.672121212	3.0836	2.202571429
New York	1535564.161	599.6	1.844923077	0.908484848	0.5996	0.428285714
Vermont	32409.32894	172.6	0.531076923	0.261515152	0.1726	0.123285714
Pennsylvania	755902.3393	701.6	2.158769231	1.063030303	0.7016	0.501142857
Ohio	652916.6188	1108.5	3.410769231	1.679545455	1.1085	0.791785714
New Jersey	598898.127	703.4	2.164307692	1.065757576	0.7034	0.502428571
Connecticut	270207.2607	517.9	1.593538462	0.78469697	0.5179	0.369928571
Massachusetts	506801.9852	503.5	1.549230769	0.762878788	0.5035	0.359642857
Tennessee	329905.1122	1863.1	5.732615385	2.822878788	1.8631	1.330785714
California	2593340.192	4790.8	14.74092308	7.258787879	4.7908	3.422

Table 8.
Data (in 2014) with the integration of HSR.

Trade Partners	(Estimated) Total Trade with Trade Partners(in millions of \$)	Time taken from the Great Montreal Area to the Trade Partners with HSR speed of (in h): tij			
		325 km/h		660 km/h	
		Adjusted Estimated Total Trade (in millions of \$)	Difference compared to the one without HSR	Adjusted Estimated Total Trade (in millions of \$)	Difference compared to the one without HSR
Newfoundland and Labrador	2199.042315	2255.127092	56.08477718	2434.622691	235.5803765
Prince Edward Island	359.853511	387.5649865	27.71147548	476.2537208	116.4002098
Nova Scotia	2300.813386	2361.742715	60.9293288	2556.742979	255.9295933
New Brunswick	3834.126706	3897.094962	62.9682563	4098.620679	264.4939731
Ontario	45392.70906	45609.30167	216.5926069	46302.4919	909.7828407
Manitoba	2362.663208	2429.778515	67.11530679	2644.57659	281.9133825
Saskatchewan	1622.714426	1690.619175	67.90474895	1907.943808	285.229382
Alberta	9045.255363	9169.48125	124.2258869	9567.057939	521.8025758

Trade Partners	**(Estimated) Total Trade with Trade Partners(in millions of $)**	**Time taken from the Great Montreal Area to the Trade Partners with HSR speed of (in h): tij**			
		325 km/h		**660 km/h**	
		Adjusted Estimated Total Trade (in millions of $)	**Difference compared to the one without HSR**	**Adjusted Estimated Total Trade (in millions of $)**	**Difference compared to the one without HSR**
British Columbia	5831.875963	5932.940157	101.0641937	6256.389388	424.5134249
Yukon	50.04212888	65.25014327	15.20801439	113.9223818	63.88025296
Northwest Territories	173.1795022	193.0160353	19.83653313	256.5015403	83.32203814
Nunavut	188.3608222	204.2602757	15.89945352	255.1454192	66.78459703
Texas	6344.752878	6599.860046	255.1071675	7416.313568	1071.56069
New York	4540.279308	4890.845128	350.5658205	6012.807723	1472.528415
Vermont	2696.44319	2788.390284	91.94709419	3082.660844	386.2176536
Pennsylvania	2574.116512	2824.75727	250.6407582	3626.916348	1052.799836
Ohio	2363.794476	2575.653476	211.8590003	3253.694116	889.8996403
New Jersey	1906.52723	2133.583774	227.0565437	2860.26314	953.7359105
Connecticut	1733.10957	1907.372346	174.262776	2465.088771	731.9792008
Massachusetts	1659.213983	1887.92553	228.7115471	2619.901625	960.6876415
Tennessee	1582.727749	1723.513192	140.7854433	2174.087641	591.3598917
California	1550.523881	1820.311415	269.7875341	2683.748476	1133.224595
Total	100312.1252	103348.3894	3036.264267	113065.7513	12753.62612
Change in %		3.02681681		12.71394271	

Table 9.
Data of great Montreal Area's trade (in 2014) with the integration of HSR.

Thus, integrating in the formula, we obtain

Through simulation, we can see an integration of HSR of 325 km/h will bring $CAN 3036.264267 million of trade in addition (+3.02681681%); 660 km/h for $CAN 12753.62612 million of trade in addition (+12.71394271%).

6. Discussion

City sustainable logistics relates to many infrastructures as well as software like authority policy, IT applications, and incorporations of both hardware and software. When HSR transportation is put into daily running, it needs to research what will happen to help city sustainable logistics and what will make bad efforts to the circulation. Normally, because of the high speed of HSR, it will advance economic or GDP development in society. However, other transportation methods may decrease their activities or even fail to the new strong challenge coming from HSR, which could beat down those professional fields, for example, the short airlines, by which people would take several hours to board, travel, and leave the airport; if by HSR, 1 hour enough to pass 400 or 600 km. Then, we need to research more what is the reality of economic multi-process by the application of both HSR and traditional vehicles, and how much the degree of their effectiveness by the process. However, when we apply more data to make AI simulation, which also verifies the correctness of the previous logical process, one of the best solutions is:

$$\mathrm{B} = e^{a_0} = \mathrm{e}^{0.01528038323988512}$$

$$a = a_1 = 0.19633948638164284$$

$$b = a_2 = 0.46481636346029886$$

$$c = a_3 = 0.2607251197660339$$

$$\mathrm{R}^2 = 0.5407492632618118$$

7. Conclusion

Through the simulation and AI analysis with multivariable linear regression with MSE and gradient descent, based on the gravitational model for trade, we can conclude an integration of a high-speed transportation vehicle such as HSR will improve the volume of trade, thus positively influencing the economy of the Great Montreal Area.

Further improvements for a better estimated total volume after the integration of HSR can be considered.

1. Although this chapter considers the most recent public data as inputs that Statistics Canada can offer, these data are the ones from 7 to 10 years ago. Because of the economic changes that happened in the Great Montreal Area, more recent data can improve the results. In addition, if one can obtain of the repartition of trade of the Great Montreal Area and its trade partners (instead of the province of Quebec and the Great Montreal Area), more precise results can be obtained. Plus, while most data are based on 2014s, data used for the repartition of transportation modes to estimate the HSR speed is 2011s which is the newest data that can be found. Thus, data used in different year may also not depict completely the economic picture.

2. Inclusion of data of additional trade partners as inputs can improve the results and give insights for the differences between the estimated and actual trade volume, leading to elements as additional inputs that one can consider including in the linear regression.

3. Because of the complexity of the multivariable linear regression, further estimation techniques can be used such as interpolation or maximum likelihood estimation to double check the simulated results.

Appendix

See **Tables A1–A4**

DOI: http://dx.doi.org/10.5772/intechopen.1005318

		Number of Simulations (m)	
		10^3	10^4
α	10^{-4}	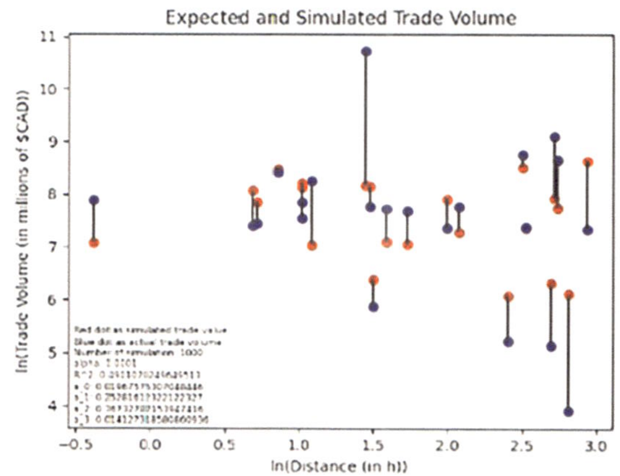	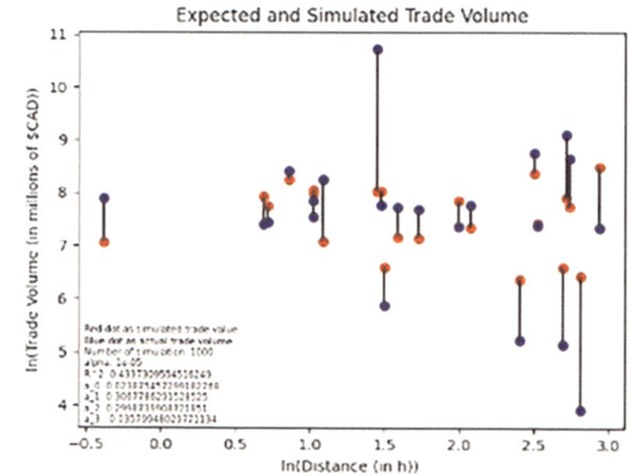
	10^{-5}	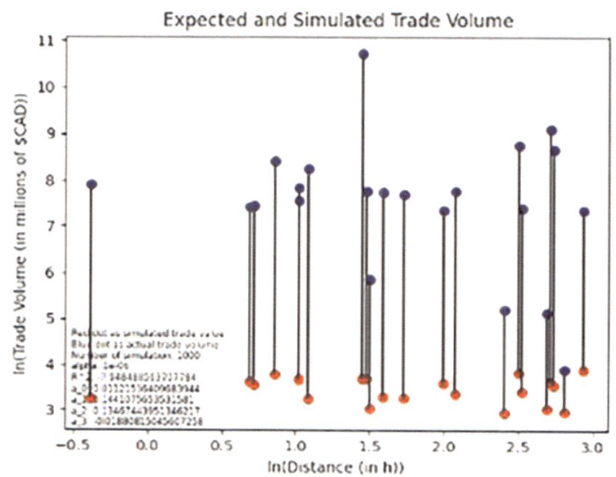	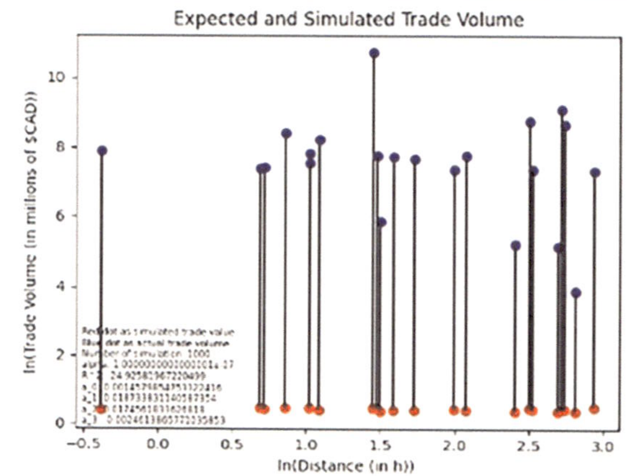

Table A1.
Simulation results.

		Number of Simulations (m)	
		10^3	10^4
α	10^{-4}	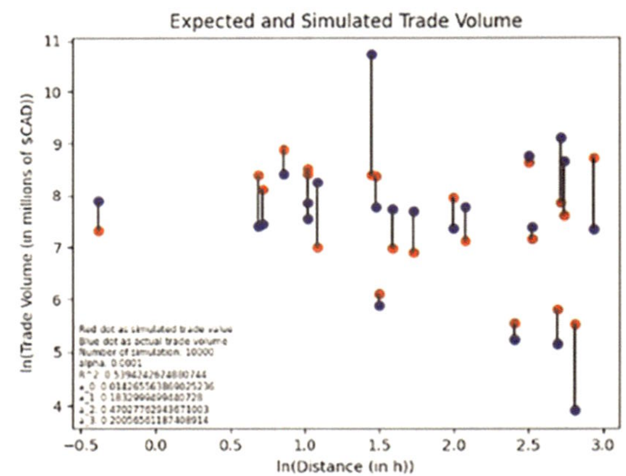	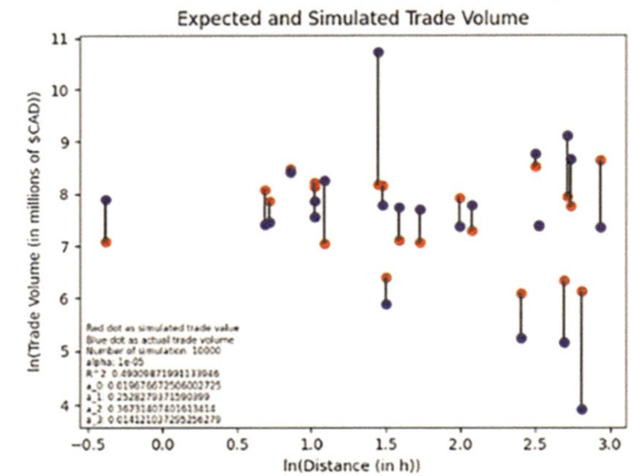
	10^{-5}	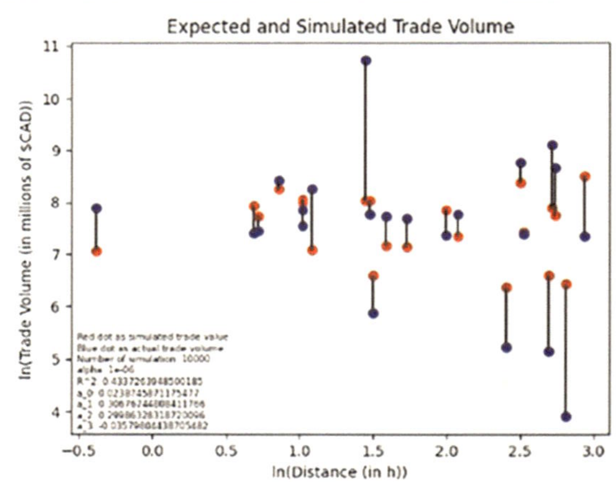	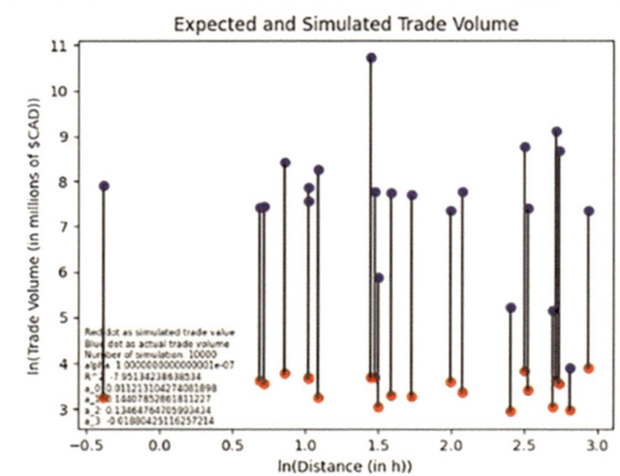

Table A2.
Simulation results.

		Number of Simulations (m)	
		10^3	10^4
α	10^{-4}	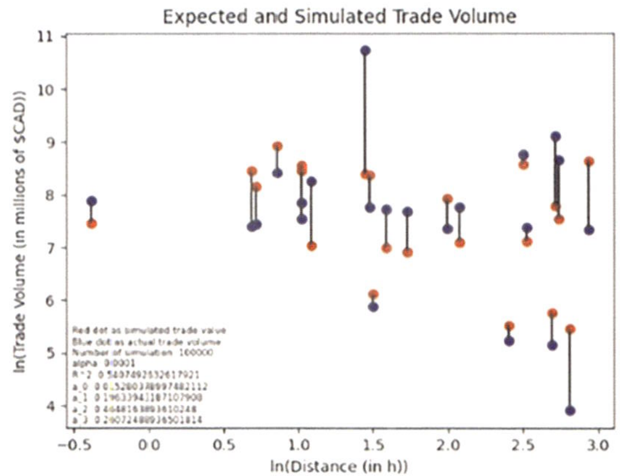	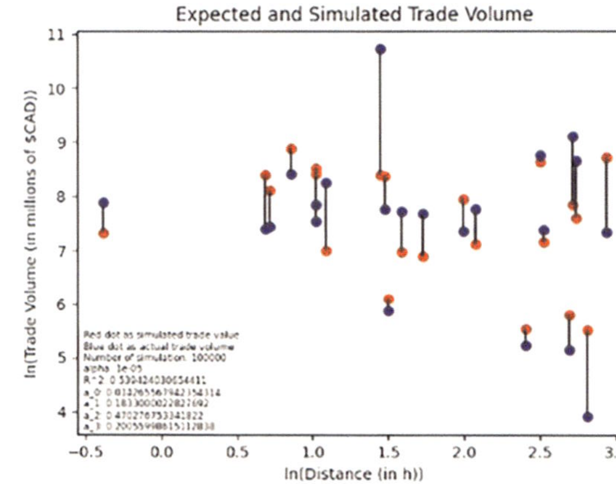
	10^{-5}	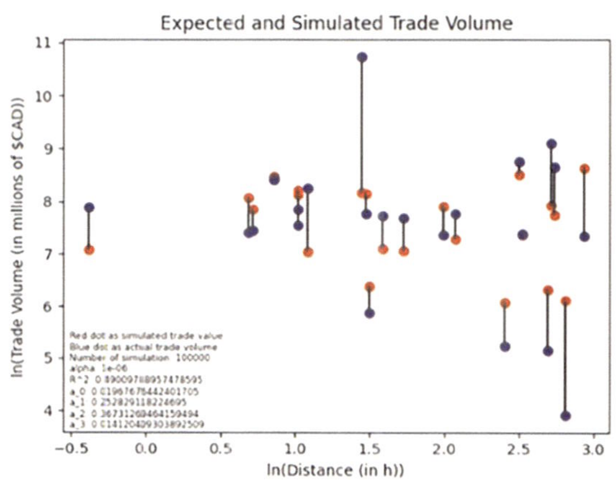	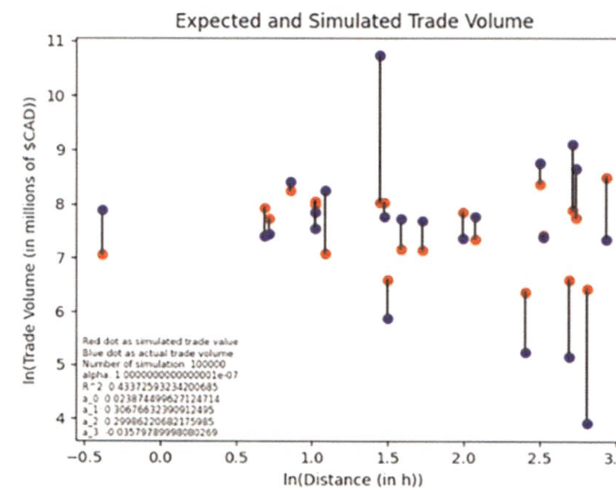

Table A3.
Simulation results.

		Number of Simulations (m)	
		10^3	10^4
α	10^{-4}	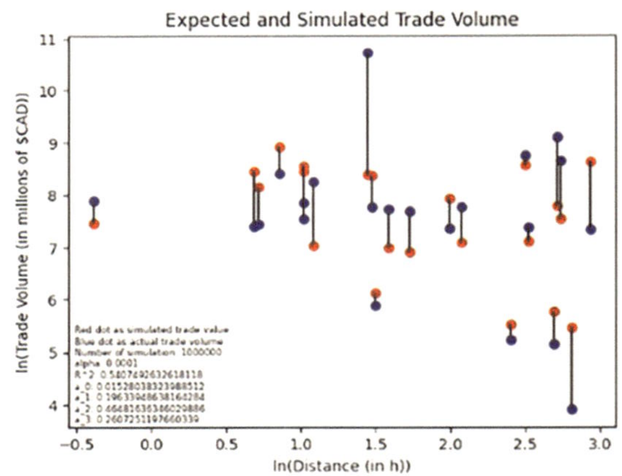	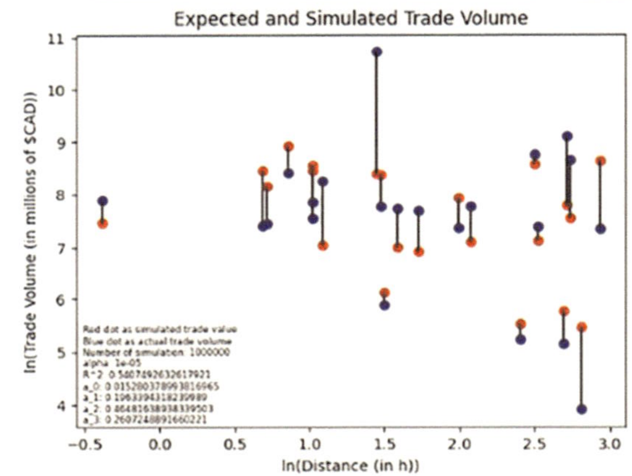
	10^{-5}	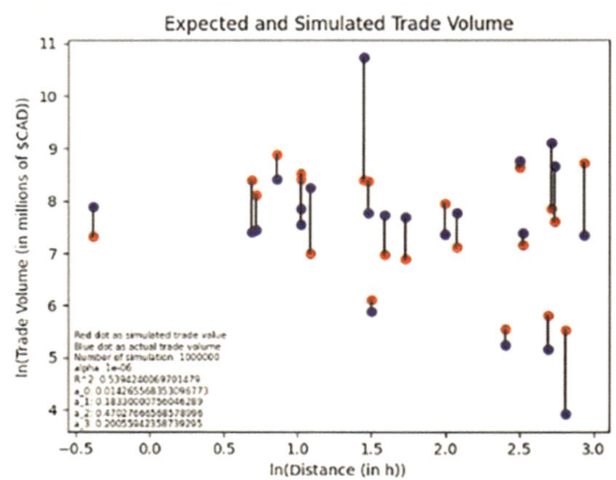	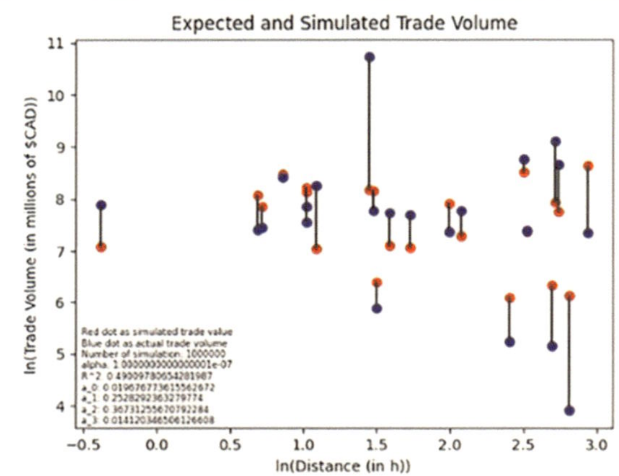

Table A4.
Simulation results.

Author details

Yonglin Ren[1*], Xinyue Ren[2] and Anjali Awasthi[1]

1 Concordia University, Montreal, Canada

2 Mathmatical Science, Mcgill University, Montreal, Canada

*Address all correspondence to: royalencrypt@gmail.com

References

[1] Li H, Dong X, Jiang Q, Dong K. Policy analysis for high-speed rail in China: Evolution, evaluation. Transport Policy. 2021;**106**:37-53

[2] Canada Federal Sustainable Development Strategy [Internet]. Government of Canada. 2016. Available from: https://www.canada.ca/en/environment-climate-change/services/sustainable-development/synthesis-report-public-consultations-2016-2019.html [Accessed: November 27, 2021]

[3] Gogna A, Tayal A. Metaheuristics: Review and application. Journal of Experimental & Theoretical Artificial Intelligence. 2013;**25**(4): 5035-5026

[4] Holland JH. Adaptation in Natural and Artificial Systems. University of Michigan Press; 1975

[5] Murthy CA, Chowdhury N. In search of optimal clusters using genetic algorithm. Pattern Recognition Letters. 1996;**17**(8):825-832

[6] Ali MM, Storey C. Aspiration based simulated annealing algorithm. Journal of Global Optimization. 1997;**11**:181-191

[7] Locatelli M. Simulated annealing algorithms for continuous global optimization. In: Pardalos PM, Romeijn HE, editors. Handbook of Global Optimization. Vol. 2. 2002. pp. 179-180

[8] Shekhawat A, Poddar P. In: Boswal D, editor. Artificial Intelligence Seminar. Indian Institute of Technology Bombay; 2009. p. 27

[9] Zhang X, Huang S, Hu Y, Zhang Y, Mahadevan S, Deng Y. Solving 0–1 knapsack problems based on amoeboid organism algorithm. Applied Mathematics and Computation. 2013; **219**:9959-9970

[10] Norrman KSA. Performance-based contracting for advanced location allocation services. International Journal of Physical Distribution & Location allocation Management. 2015;**45**(6): 592-617. DOI: 10.1108/IJPDLM-11-2014-0267

[11] Lu M, Borbon-Galvez Y. Advanced location allocation and supply chain management for intelligent and sustainable transport. In: 19th ITS World Congress. 2012

[12] Takai E. The Role of Operations Research towards Advanced Location Allocation. NISTEP Science & Technology Foresight Center; 2009

[13] Issaoui Y et al. An advanced system to enhance and optimize delivery operations in a smart location allocation environment. IEEE Access. 2022;**10**: 6175-6193

[14] Shinohara M. Paradigm shift in maritime transport. The Asian Journal of Shipping and Location allocation. 2009; **25**(1):57-67

[15] Miyashita K. Structural change in the international advanced location allocation. The Asian Journal of Shipping and Location allocation. 2009;**25**(1): 121-138

[16] Kaszubowski D. Recommendations for urban freight policy development in Gdynia. Transportation Research Procedia. 2012;**12**:886-899

[17] Anderson JE. The Gravity Model, NBER Working Paper Series. National

Bureau of Economic Research. pp. 1-3. Available from: https://www.nber.org/system/files /working_papers/w16576/w16576.pdf

[18] Canadian Trade Balances [Internet]. Government of Canada; 2020. Available from: https://www.ic.gc.ca/app/scr/tdst/tdo/crtr.html?grouped=GROUPED & searchType=All&naArea=P24&countryList=SALL&toFromCountry=CDN&reportType=TB&timePeriod=10%7CComplete+Years&currency=US&productType=HS6&runReport=true [Accessed: November 27, 2021]

[19] Gross domestic product (GDP) at basic prices, by sector and industry, provincial and territorial [Internet]. Government of Canada; 2020. Available from: https://www150.statcan.gc.ca/t1/tbl1/en/tv.action?pid=3610048701 [Accessed: November 27, 2021]

[20] Gross Domestic Product (GDP) by state, 1997-2016 [Internet]. U.S. Bureau of Economic Analysis (BEA). ArcGIS Hub; 2018. Available from: https://hub.arcgis.com/datasets/a2a8826c4ba44f89a684000fe7c05b8c_0?geometry=-126.256%2C20.471%2C1.185%2C63.314&selectedAttribute=GDP_by_State___97_to_16__dol_39 [Accessed: November 27, 2021]

[21] Yearly Average Rates [Internet]. OFX; 2021. Available from: https://www.ofx.com/en-ca/forex-news/historical-exchange-rates/yearly-average-rates/ [Accessed: November 27, 2021]

[22] Google Maps [Internet]. Google; 2021. Available from: https://www.google.com/maps [Accessed: November 27, 2021]

[23] Gross domestic product of Montréal, Québec, at constant prices from 2009 to 2017 [Internet]. Statista; 2021. Available from: https://www.statista.com/statistics/652749/real-gdp-of-montreal-quebec/ [Accessed: November 27, 2021]

[24] Gross domestic product of Quebec, Canada from 2000 to 2019 [Internet]. Statista; 2021. Available from: https://www.statista.com/statistics/577535/gdp-of-quebec-canada/ [Accessed: November 27, 2021]

[25] Industry profile [Internet]. Government of Canada; 2013. Available from: https://www.ic.gc.ca/eic/site/dsib-logi.nsf/eng/h_pj00541.html [Accessed: November 27, 2021]

[26] April 18, 2011 Average Truck Speeds [Internet]. Office of Energy Efficiency & Renewable Energy; 2021. Available from: https://www.energy.gov/eere/vehicles/fact-671-april-18-2011-average-truckspeeds#:~:text=The%20average%20speed%20of%20trucks,crossings%2C%20and%20in%20mountainous%20terrain [Accessed: November 27, 2021]

[27] Morris Hugh. The fastest passenger plane in the sky? It might surprise you. The Telegraph; 2017. Available from: https://www.telegraph.co.uk/travel/travel-truths/which-passenger jet plane flies-thefastest/#:~:text=Average%20cruising%20speed%20for%20a, tickled%202%2C193 mph%20in%201976.

[28] Canadian Pacific's average train speed from FY 2013 to FY 2020 [Internet]. Statista; 2021. Available from: https://www.statista.com/statistics/548553/average-train-speed-canadian-pacific-railway/ [Accessed: November 27, 2021]

[29] Transportation Modes Revealed: Water [Internet]. Union Pacific; 2019. Available from: https://www.up.com/customers/track-record/tr082719-water-

pros-cons.htm#:~:text=Ships%20and%20barges%20typically%20travel,to%20arrive%20at%20their%20destination [Accessed: November 27, 2021]

[30] Richard N. Fact sheet. High speed rail development worldwide. In: Transportation. EESI Environmental and Energy Study Institute; 2018. Available from: https://www.eesi.org/papers/view/fact-sheet-high-speed-rail-development-worldwide#:~:text=High%2Dspeed%20trains%20can%20generally,(120%E2%80%93160%20mph)

[31] Yile A, Hongqi L, Liping Z, Sikandar A, Zhongguo Y. The linear random forest algorithm and its advantages in machine learning assisted logging regression modeling. Journal of Petroleum Science and Engineering. 2019;**174**:776-789. Available from: https://reader.elsevier.com/reader/sd/pii/S0920410518310635?token=1957CF0937CFA4ED6BE9D212F0182C9B4EFCDB838760CC9BB194CD892A8AB283205F83661ACE4D628F5BAAD2AB5C0C33&originRegion=us-east-1&originCreation=20210829135138

[32] Why the gradient is the direction of steepest ascent [Internet]. Khan Academy; 2021. Available from: https://www.khanacademy.org/math/multivariable-calculus/multivariable-derivatives/gradient-and-directional-derivatives/v/why-the-gradient-is-the-direction-of-steepest-ascent [Accessed: November 27, 2021]

Chapter 5

Application of Machine Learning for Predictive and Prognostic Reliability in Flexible Shop Floor

Ayoub Chakroun and Nidhal Rezg

Abstract

Flexible workshops are essential components of modern industry, enabling flexible and efficient production. However, to ensure their proper functioning and prevent unexpected breakdowns, it is crucial to monitor their reliability. Production stoppages caused by unforeseen breakdowns can lead to significant financial losses. This chapter proposes to explore the use of Machine Learning (ML) for predicting the reliability of flexible workshops, thus identifying dates for Preventive Maintenance (PM) interventions and optimizing production management. The objectives of this exploration include the presentation of new predictive model developments and the description of ML models capable of predicting workshop reliability based on real-time data, such as equipment monitoring, production data, and maintenance histories. It also aims to identify optimal times for PM interventions, minimizing production disruptions and optimizing resource utilization. Additionally, the chapter will propose cost optimization models to prevent unplanned breakdowns, extend equipment lifespan, optimize spare parts usage, and maximize productivity by avoiding production interruptions and ensuring the smooth operation of the flexible workshop.

Keywords: machine learning, predictive maintenance, prognostic reliability, health assessment, flexible workshop

1. Introduction

In today's economic landscape, characterized by intense globalization and progressively exacting markets, industries are compelled to enhance the efficacy and efficiency of their production lines. This drive aims to bolster their competitiveness and meet the escalating demands of their clientele. The amalgamation of connectivity, data utilization, emerging technologies, inventory streamlining, personalized manufacturing, and regulated production has ushered in the era known as Industry 4.0, which appears to be an unstoppable force shaping the industrial realm.

Industry 4.0 has ushered in a new era of advanced technologies including Artificial Intelligence (AI), Internet of Things (IoT), Cyber-Physical Production Systems (CPPS), and Big Data and Analytics [1–3]. This technological evolution is transforming processes, expertise, business models, and the maintenance of manufacturing operations. Maintenance, crucial in ensuring equipment functions effectively, has become more vital due to the high value of manufacturing assets and their propensity to degrade over time. The maintenance department's reliability is critical in boosting productivity. As noted by Parida and Chattopadhyay [4], evaluating maintenance performance is increasingly essential. Several studies emphasize maintenance's role in enhancing efficiency and economic sustainability in manufacturing [5–7]. Maintenance is recognized as a key driver for competitive advantage in the industry [8].

Recent advancements in technology, especially in smart resources, have facilitated real-time data collection and integration, aiming to improve reliability, lower repair costs, and predict equipment degradation. Predictive maintenance, in particular, is pivotal in reducing unexpected downtime in manufacturing. Research in this field investigates the impact of Big Data and Machine Learning (ML) techniques on industrial maintenance [9]. AI and ML are extensively used to monitor manufacturing processes and forecast downtimes. Leukel et al. [10] explore ML integration for predicting machine breakdowns, proposing a decision-support framework to minimize downtime in CPPS [11]. Chaudhuri [12] suggests a Hierarchical Modified Fuzzy Support Vector Machine (HMFSVM) for predicting vehicle failure trends.

Kamariotis et al. [13] focus on Prognostic Health Management (PHM) to predict the Remaining Useful Life (RUL) of equipment, introducing a metric for evaluating data-driven prognostic algorithms in Predictive Maintenance (PdM) decisions. Similarly, Khazaelpour and Hashemkhani Zolfani demonstrate the effectiveness of an AI-based predictive maintenance model, the FUCOM-ANN hybrid approach, which outperforms standard Artificial Neuronal Network (ANN) predictions in reliability and cost optimization [14].

The literature also presents various predictive maintenance models suited to different constraints in health assessment contexts. Elkateb et al. [15] describe using AdaBoost ML algorithms for real-time classification of machine stops in knitting machines, achieving a 92% accuracy rate. Additionally, a multi-agent framework, SCMEP, is introduced for integrating production and predictive maintenance scheduling [16]. Zonta et al. [17] propose a model using deep neural networks for optimizing production scheduling. One significant limitation of existing predictive models is the absence of benchmark datasets and standardized evaluation metrics. The presence of such benchmark datasets and standardized evaluation metrics is pivotal to ensuring impartial and consistent comparisons among various predictive maintenance models.

However, the literature also identifies gaps, notably the lack of comprehensive comparisons among various ML algorithms in PdM. Studies often focus on a single algorithm without robust comparisons to others. A holistic evaluation of different algorithms would provide insights into their respective advantages and challenges in various PdM scenarios.

Real-world PdM implementation faces complexities such as varying operating conditions and data inconsistencies. Unfortunately, some studies overlook these challenges in ML model design and evaluation. This study aims to address these real-world challenges to enhance the effectiveness and robustness of PdM models. A significant limitation in current predictive models is the lack of benchmark datasets and standardized evaluation metrics, which are crucial for fair and consistent comparisons across different PdM models.

The rest of this paper is organized as follows. Section 2 introduces the problem statement. In Section 3, we elaborate on the materials and methods, including the proposed method, data analysis and processing procedures, and the developed predictive models using ML techniques. Section 4 presents the results and discussions. Finally, Section 5 offers conclusions and perspectives.

2. Problem statement

This chapter revolves around the operational health of an automated robot responsible for packaging mechanical products in a flexible manufacturing system. **Figure 1** depicts a computer-based simulation model of the dedicated shop floor. This latter is designed to efficiently perform product loads with few settings independently of the production line. However, the factory is grappling with production capacity challenges due to difficulties in activating the conditioning unit. Consequently, the article explores a case study centered on the packaging robot, specifically predicting the degradation of its power transmitters under demanding mechanical and thermal conditions, aiming to identify optimal times for preventive maintenance interventions. Notably, the conditioning robot is equipped with specialized sensors and advanced technology and connected to a local area network, transforming the workshop into a CPPS. These probes measure parameters affected by changes and interruptions in the product packaging process.

3. Materials and methods

3.1 Suggested method

This paper delves into a practical case study involving a factory specializing in mechanical products manufacturing. Our main emphasis is on deploying PdM for a robot responsible for packaging final products, particularly forecasting the deterioration of its power transmitter. Our goal is to introduce and compare two predictive maintenance models within the framework of smart era. To accomplish this, we examine and discuss the importance of failure prognosis. Consequently, we develop two predictive models using supervised ML techniques and another one using unsupervised ML. These models forecast the gradual decline of the robot and

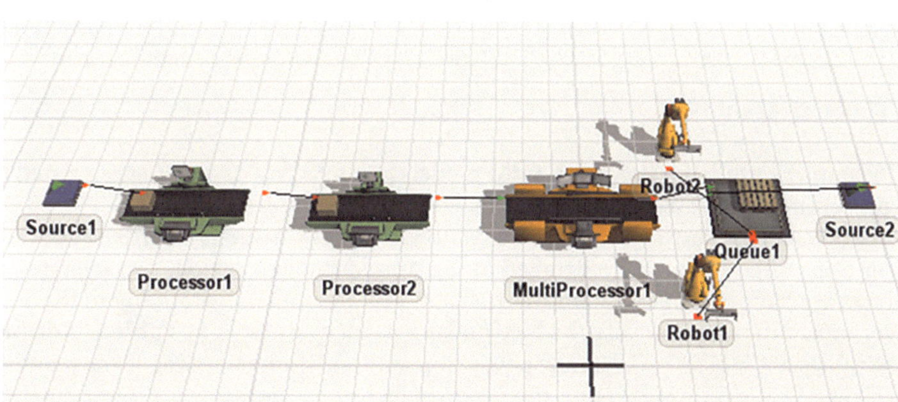

Figure 1.
Flexible shop floor facility layer [3].

anticipate its future conditions, aiding maintenance personnel in making well-informed decisions regarding PM interventions. Following this, we will offer the rationale behind the selected predictive models and conduct an in-depth comparative analysis between them.

The initial predictive model relies on a discrete Bayesian filter (DBF), which has been identified as highly suitable for addressing this specific problem [18–20]. A primary strength of the DBF lies in its adeptness at integrating information from various sources such as processes, configuration variables, and sensor data. Moreover, the DBF effectively manages uncertainties inherent in noisy processes and sensor measurements. Its adaptability to assimilate new information and respond to changing operational conditions allows for swift real-time decision-making within short operating periods. In the context of Industry 4.0, the DBF is well-matched for predictive maintenance, which is crucial for prompt issue identification and suitable corrective measures.

The second model suggested is a predictive strategy based on the Naïve Bayes Filter (NBF), a probabilistic classification model extensively used in machine learning. The NBF exhibits various advantages that render it appropriate for our context. These advantages encompass computational efficiency, adeptness in managing large datasets, resilience to irrelevant features, and ease of implementation. In summary, the Naïve Bayes filter emerges as a practical option for predictive maintenance in our scenario.

Furthermore, an additional model was established utilizing Isolation Forest, an unsupervised learning technique. This approach is particularly effective in identifying anomalies or outliers within a dataset. By leveraging the Isolation Forest method, this model excels in detecting deviations or irregularities that may signify potential issues or anomalies within the operational system.

The primary objectives of this paper are delineated as follows:

- Conducted analysis and structuring of massive data

- Crafting two distinct predictive maintenance models utilizing supervised ML techniques, namely DBF and NBF, and another model based on unsupervised ML: Isolation forest.

The flowchart representing the proposed method in this paper is presented in **Figure 2**.

3.2 Theoretical study

In simpler terms, the suggested PdM models focus on forecasting issues related to the power transmitter of a conditioning robot, as shown in **Figure 3**.

The primary aim is to scrutinize and comprehend the distinctive attributes of these power transmitters, which are detailed in the subsequent subsection. To facilitate analysis, we have compiled a comprehensive list of parameters pertaining to the power transmitter, specifically the belt-pulley system, presented in **Table 1**. Notably, these values are indicative of a new power transmitter. Understanding these parameters allows us to gain insights into the functionality of the power transmitter, analyze extensive data, and detect any deviations from normal values during the predictive maintenance process.

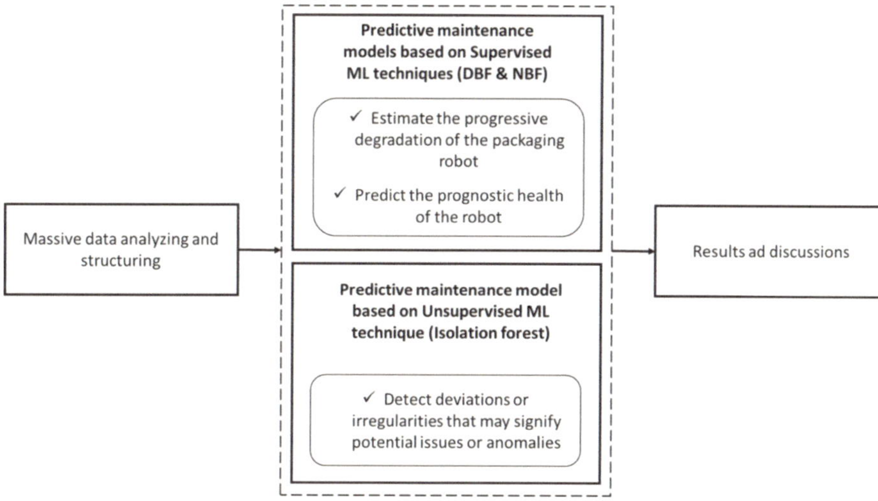

Figure 2.
Proposed method's flowchart.

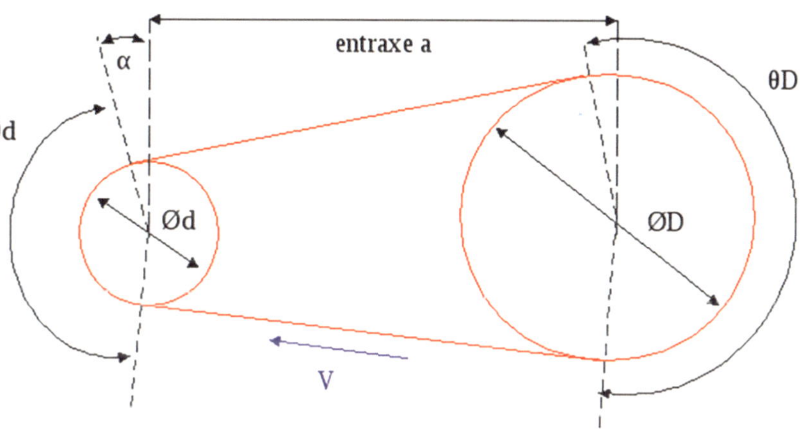

Figure 3.
Packaging Robot's power transmitter.

Indeed, the predictive maintenance models developed within the presented research offer insight into the health and performance of the power transmitter in the packaging robot. These models are tailored to oversee and uphold the power transmitter functionality within the packaging robot, utilizing sophisticated algorithms and ML techniques to anticipate and detect potential issues before they escalate into equipment failures. Through the implementation of these models, our objective is to enhance the reliability and efficiency of the packaging robot, enabling us to identify optimal times for PM interventions, minimizing production disruptions, and optimizing resource utilization.

In simpler terms, the packaging robot incorporates a dedicated set of smart and embedded sensors, which are interconnected via a local area network, transforming the workshop into a CPPS. The most important function involves real-time measurement of parameters susceptible to alterations and disruptions in the conditioning

Parameter	Signification
N_D	Large pulley speed
N_d	Small pulley speed
W_D	Large pulley angular speed
W_d	Small pulley angular speed
D	Large pulley diameter
d	Small pulley diameter
C	Torque
a	Interaxis
V	Speed
θ_D	Large pulley winding angle
θ_d	Small pulley winding angle
L	Belt length
T	Stretched strand tension
t	Soft strand tension
T_0	Initial tension
f	Friction Coefficient
P	Admissible power

Table 1.
Related parameters of the robot's power transmitter.

process. Throughout the packaging operation, these sensors collect data, which is securely transmitted to a centralized server.

In the subsequent section, our aim is to process the sensor-derived data into a format suitable for storage in a Database. Subsequently, we will employ ML techniques to construct predictive models. These models will evaluate the current degradation status of the power transmitter in the robot and forecast its future performance. Access to real-time degradation updates and future behavior predictions allows us to create a comprehensive maintenance plan, enhancing the reliability and efficiency of our maintenance interventions.

It is crucial to emphasize that our research is specifically dedicated to implementing predictive maintenance strategies for the conditioning robot, with a focal point on its power transmission system.

The following section delves into the specifics of the data examined.

3.3 Process of analyzing data

3.3.1 Corresponding data

In this section, our objective is to analyze a data frame covering the period from 2020 to 2022, including records of 2900 packaging processes conducted during this timeframe. These measurements are stored in files, each containing values for 15 parameters (as outlined in **Table 1**). Due to the substantial volume of data, with 43 k

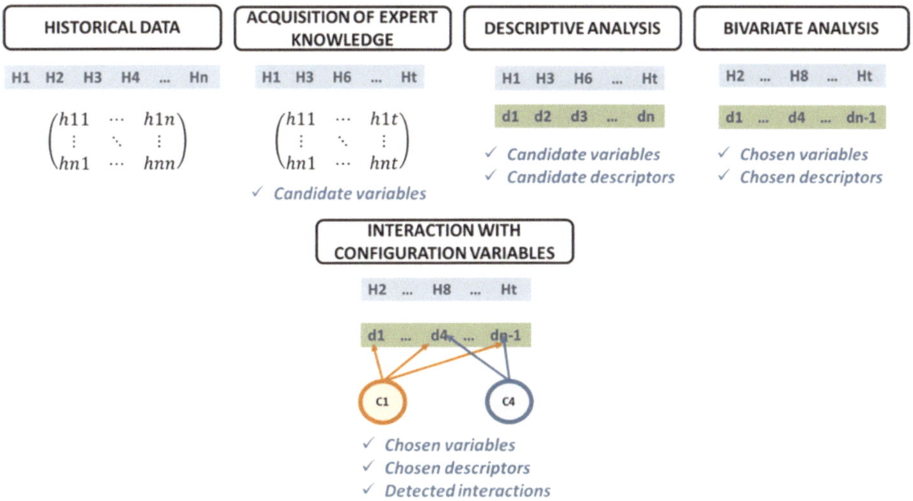

Figure 4.
Flowchart depicting the successive steps of data analysis designed to extract the most relevant descriptors summarizing the raw data [20].

variables, it is essential to organize this information efficiently. To achieve this, we will use representative descriptors, elaborated upon in the following sections.

By referring to the literature, Ruiz-Sarmiento, and his colleagues have introduced a descriptor as a concise representation of specific data elements that elucidates and retains information essential for addressing a particular problem [18] and will contain information concerning the conditioning robot's health for our case study.

The forthcoming data analysis will proceed through four distinct steps aimed at identifying pertinent descriptors capable of effectively summarizing data and offering insights for constructing predictive models for the packaging robot's power transmitter. The entire data analysis process is depicted in **Figure 4**.

3.3.2 Acquisition of expert knowledge

To ensure a thorough and precise analysis of collected data, thereby avoiding arbitrary or inaccurate methods [21], it is essential to seek input from plant experts. Their expertise is particularly valuable in the conditioning process, where we utilized human elicitation techniques with production floor specialists. Their input was crucial for identifying relevant descriptors and effectively analyzing the collected data.

The following points provide more detail on the information gathered:

- *Identification of Key Variables*: Of the 15 variables identified in each packaging process, two are related to process progression: the operating range and the conditioning robot's capacity. The remaining 11 variables result from the configuration of the process and include elements like the speed, angular speed, and linear speed of the small pulley, as well as the winding angle. The last two variables – the coefficient of friction and the tension of the strand – depend on the condition of the power transmitter and could act as indicators of its health. These are integrated into the corresponding descriptor.

- *Exploring Potential Interactions*: Experts highlighted the importance of examining potential interactions between variables. This exploration forms the fourth step in our proposed data analysis methodology.

- *Determining Factors Affecting Asset Deterioration*: Plant experts identified key factors indicative of the power transmitter's deterioration. These are closely linked to configuration variables such as the capacity of the packaging robot and the speed of the small pulley, and are considered crucial for estimating the degradation state of the asset.

3.3.3 Descriptive analysis

In the second phase, our objective is to scrutinize the conduct of variables throughout the packaging procedures of the final products and to pinpoint characteristic descriptors suitable for their delineation. We scrutinized various variables linked to the robot's transmitter, as highlighted in **Table 1**.

In a typical lifespan, a newly installed robot's transmitter exhibits functionality for approximately 40 product-conditioning tasks. Upon meticulous examination of **Figure 5**, the initiation of power transmitter degradation became evident around process number 35, primarily attributable to substantial variations in input variables. This inference underscores the notable impact of the quantity of packaging processes on the health condition of the power transmitter.

3.3.4 Bivariate analysis

In the third phase, our aim is to construct models that establish a relationship between the patterns manifested by the parameters, saved in V, and the deterioration status of the robot's transmitter, with a particular focus on the belt component. The objective is to discern vectors that precisely portray the condition of the asset.

Through this sub-section, we aim to identify appropriate descriptors for our PdM framework; we employed the coefficient of determination, represented as R^2. This measure assesses the accuracy of forecasts generated by the linear regression model. In

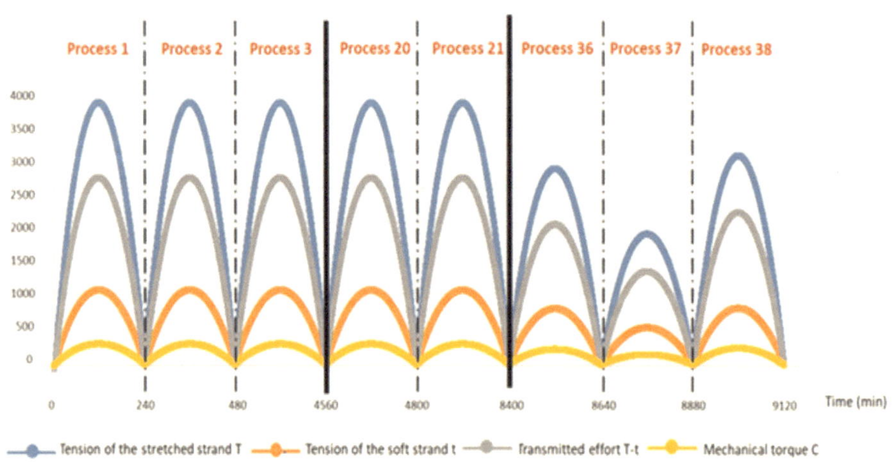

Figure 5.
Graphical representation of critical parameters [20].

V1	V2	V3
Small pulley speed	Large pulley speed	Robot's capacity
Small pulley winding angle	Large pulley winding angle	Packaging process
Small pulley angular speed	Large pulley angular speed	
Soft strand tension	Stretched strand tension	
Friction coefficient		

Table 2.
Potential descriptors.

order to assess the predictive performance of our models (DBF and NBF), we conducted a linear regression and examined its residuals, obtaining R^2.

Following established literature, descriptors exhibiting R^2 values close to 1 were selected, suggesting a strong correlation among variables and an increased probability of providing precise predictions within PdM frameworks.

Table 2 highlights the corresponding descriptors (V1, V2, and V3).

3.4 DBF based on machine learning techniques

The chosen model for PdM is a DBF. Following the processing and transformation of the data into a useful format, ML techniques are applied to construct the predictive model. DBF serves as a robust mechanism to assimilate uncertain and fluctuating information into a production system. In our case, we focus on a packaging robot fitted with sensors that generate imprecise measurements. Our aim is to derive the most precise estimation possible of the current condition of the power transmitter.

The chosen model for this investigation, the DBF, gauges the degradation status of the power transmitter through discrete value ranges. This model incorporates considerations of confidence in the outcomes, enhancing the accuracy of estimation.

In the following, we will design the PdM mathematical model. Through this model, we note:

- N_s: Number of values or states in which transmitter health/degradation is discretized;
- x: Discrete random variable representing the degradation of the power transmitter, $x \in \{1, \ldots, N_s\}$;
- k: Instant of time, so x_k indicates the degradation's state of the power transmitter at instant k;
- z_k: Sensor measurements available at time k; in fact it is a reference to the vector of descriptors V;
- u_k: Action taken at the moment k during the spherical bushels' assembly process;
- c_k: Process configuration parameters that influence sensor measurements.

At each instance, denoted as k, upon completion of a packaging process, we can determine the likelihood of the power transmitter experiencing a specific level of degradation after finalizing the packaging process for 100 items. This computation can be carried out using Eq. (1) as follows (where η denotes a normalization constant):

$$Bel(x_k) = \square.P(z_k|x_k, c_k, z_{k-w:k-1}) \sum_{i=1}^{Ns} P(x_k|x_{k-1})Bel(x_{k-1}) \tag{1}$$

3.4.1 Degradation's estimation

As presented below, Algorithm 1 is utilized for predicting the degradation levels of packaging robot at a specific time denoted k. During the prediction phase (lines 2 to 3), the algorithm anticipates the degradation state at time k based on the preceding belief (at time k − 1) and the action model. Following the prediction, the algorithm advances to the update phase (lines 4 to 9), where it adjusts the prior prediction using sensor measurements from the sensor model. This phase calculates the most probable degradation state aligned with the sensor measurements. Ultimately, after normalization, the belief regarding the degradation state Bel(x_k) can be determined (line 11). This algorithm, employing the Discrete Bayesian Filter (DBF) approach, integrates uncertain and variable information and inspections into the production system. By utilizing a discrete spectrum of values to represent the transmitter's health, the DBF model computes degradation state while considering confidence in the outcomes.

Algorithm 1: Discrete Bayes Filter for the estimation of the degradation's state

1. Procedure **DBF − ESTIMATION**$(\mathrm{Bel}(x_{k-1}), u_k, z_{k-w:k-1}, c_k, N_s)$

    $\mathrm{Bel}(x_{k-1})$ % Belief at the moment $x_{k\text{-}1}$

    u_k % Control action taken

    $z_{k-w:k-1}$ % Last w measurement (sensors)

    c_k % Configuration settings of the assembly process

    N_s % number of possible discretized degradation's states

2. **For** each x_k **do**

3. **For** i from 1 to N_s **do**

4. $\mathrm{Bel}(x_k) = \sum_{\mathrm{i}=1}^{\mathrm{Ns}} \mathrm{P}(x_k \mid u_k, x_{k-1}^i * \mathrm{Bel}(x_{k-1}^i)$ ***Prediction phase***

5. **End**

6. $\eta = 0$

7. $\mathrm{Bel}(x_k) = \mathrm{P}(z_k \mid x_k,\ c_k, z_{k-w:k-1} * \mathrm{Bel}(x_k)$ ***Update phase***

8. $\eta = \eta + \mathrm{Bel}(x_k)$
9. $\mathrm{Bel}(x_k) = \eta^{-1} * \mathrm{Bel}(x_k)$

10. **End**

11. **Return**(Bel(x_k))

We outline several initial conditions necessary for the development of the PdM:

- Bel (x_1 = 1) = 1: the power transmitters are new;
- Bel (x_i = 1) $\neq$ 1 = 0, whatever i $\neq$ 1 we will not be able to have a first-order degradation state.

4. Results and discussion

4.1 Estimation of the degradation state

Figure 6 provides a visual representation of the degradation progression of a power transmitter after 29 packaging processes, equivalent to 29,000 items being processed.

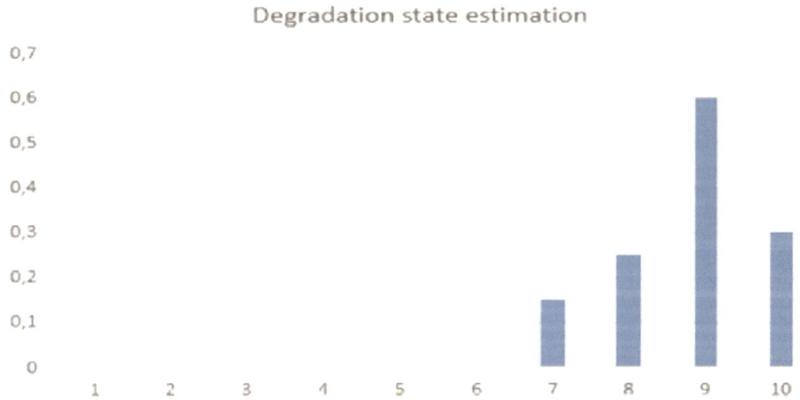

Figure 6.
Degradation states that the estimation of a power transmitter has performed 29 conditioning processes.

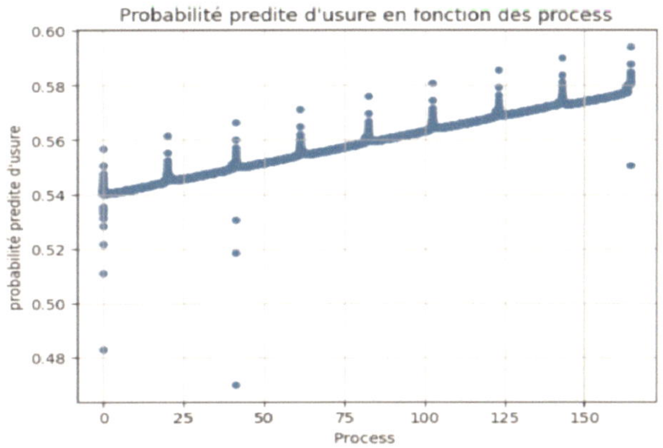

Figure 7.
Predicted wear probability of the DBF predictive model in function of process' number [3].

Figure 7 showcases the predicted wear probability attributed by the DBF predictive model concerning the process number. Notably, the graph demonstrates a consistent linear rise in wear probability as the number of processes increases.

4.2 Future state prediction

To predict the future state, incorporating additional control actions for packaging processes $u_{k+1} \ldots u_{k+4}$ (scenario 1) and $u_{k+1} \ldots u_{k+15}$ (scenario 2) is crucial for enhancing the model. Subsequently, the model is executed to generate predictions. **Figure 8** depicts the model's outcomes in two distinct scenarios: the first one is performed after 29 packaging processes with an additional 4 packaging tasks, and the second one follows 29 processes accompanied by 15 additional conditioning tasks. In the first case, maintenance personnel can deduce that the robot's power transmitter maintains a sufficient state to function after completing 33 tasks (29 + 4). Conversely, case two indicates the necessity of a maintenance task to preempt potential failures.

Comparing the results between the DBF predictive model and the NBF model rooted in machine learning, we have gauged their matching scores, reflecting their respective abilities to predict outcomes based on input data. The DBF model displays a commendable 62% score, while the NBF model trails slightly behind with a score of 55%.

Based on the outcomes, **Figures 9** and **10** derived from the Isolation Forest unsupervised learning algorithm, it is noticeable that after every 20 packaging

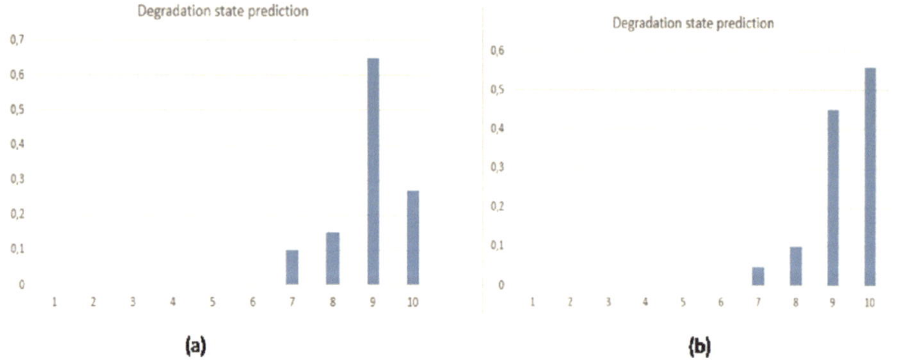

Figure 8.
Predictive outcomes from the model in two distinct scenarios [3].

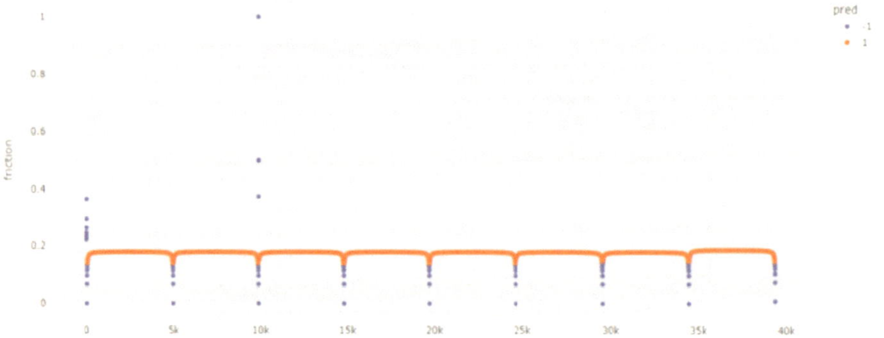

Figure 9.
Anomaly detection derived from the isolation Forest unsupervised ML technique.

DOI: http://dx.doi.org/10.5772/intechopen.1004999

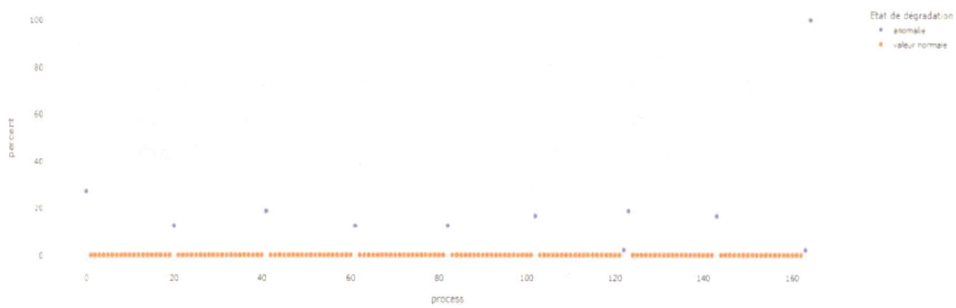

Figure 10.
Percentage of anomalies in each packaging process.

processes, the degradation state of the system exhibits a variation, fluctuating within a range of 2–27%. This variation pattern suggests a diverse level of degradation at different intervals, implying a non-linear or irregular degradation progression over these specific time intervals.

The predictions of the model are either 1 or − 1; where −1 indicates outliers or anomalous values, while 1 signifies normal values.

Conclusively, we affirm that the DBF model exhibits superior predictive capabilities compared to the NBF model. Additionally, in comparison with the Isolation Forest, the DBF model highlights enhanced predictive abilities.

5. Conclusions

In summary, adopting a Machine Learning (ML) based predictive maintenance (PdM) model aligns with Industry 4.0's objectives, marking a significant step toward its standards. This study utilized a Discrete Bayesian Filter (DBF) to evaluate the degradation status of a power transmitter in a packaging robot. When compared to a traditional ML-based model (NBF) and an Isolation Forest Unsupervised ML technique, the DBF model demonstrated enhanced predictive accuracy, achieving a superior matching score. The selection of descriptors was carried out through a rigorous coefficient of determination R^2 analysis, and predictions were facilitated by a linear regression model. The DBF model excelled in incorporating uncertain and noisy sensor data, accurately estimating the transmitter's degradation state. This model is set to support maintenance teams in developing effective strategies, reducing downtime, and increasing productivity in the production system. Its effectiveness in streamlining maintenance planning and cost-efficiency aligns with Industry 4.0's fundamental goals. A key aim of smart factories is to implement PdM using AI technologies to prevent equipment failures, which is precisely what this study's PdM model achieves by leveraging historical sensor data and ML algorithms.

Our research necessitates a thorough examination of its implications on the overall efficiency of a flexible shop floor. The findings focus on PdM models for a packaging robot's power transmitter. On the one hand, PdM models introduce a proactive maintenance approach. By predicting issues in the power transmitter beforehand, maintenance can be scheduled strategically, minimizing unplanned downtimes and optimizing resource use, thereby enhancing the shop floor's workflow.

However, the potential challenges and limitations must be critically assessed. Predictive models rely on accurate data and consistent system dynamics. Real-world complexities like variable operating conditions and unexpected anomalies could impact model accuracy. Thus, while our research marks a valuable advancement toward efficiency, it also underscores the necessity for continual model refinement and adaptability to changing conditions on the shop floor.

To conclude, our research offers promising methods to improve efficiency on flexible shop floors or other manufacturing systems facility layouts through PdM, but it requires careful consideration of both its advantages and limitations. Recognizing these aspects ensures a comprehensive understanding of how our findings can be effectively implemented and adapted for practical use.

Looking forward, our aim is to compare our highly predictive model (DBF) with a preventive maintenance model based on expert knowledge, highlighting the contrasts in predictive power and practical application.

Conflict of interest

The authors declare no conflict of interest.

Author details

Ayoub Chakroun* and Nidhal Rezg
Laboratory of Computer Engineering, Production, and Maintenance, University of Lorraine, Nancy, France

*Address all correspondence to: ayoub.chakroun@univ-lorraine.fr

References

[1] Chakroun A, Hani Y, Elmhamedi A, Masmoudi F. A proposed integrated manufacturing system of a workshop producing brass accessories in the context of industry 4.0. International Journal of Advanced Manufacturing Technology. 2022;**127**:2017-2033. DOI: 10.1007/s00170-022-10057-x 2

[2] Chakroun A, Hani Y, Masmoudi F, El Mhamedi A. Digital transformation process of a mechanical parts production workshop to fulfil the requirements of industry 4.0. In: LOGISTIQUA 2022 IEEE: 14th International Conference of Logistics and Supply Chain Management LOGISTIQUA 2022–25–27 May 2022, ELJADIDA, Morocco. 2022. p. 6. DOI: 10.1109/LOGISTIQUA55056.2022.9938099

[3] Chakroun A, Hani Y, Turki S, Rezg N, Elmhamedi A. Development of predictive maintenance models for a packaging robot based on machine learning. In: Advances in Production Management Systems. Production Management Systems for Responsible Manufacturing, Service, and Logistics Futures. APMS 2023. IFIP Advances in Information and Communication Technology. Vol. 690. Cham: Springer; 2023. DOI: 10.1007/978-3-031-43666-6_46

[4] Parida A, Chattopadhyay G. Development of a multi-criteria hierarchical framework for maintenance performance measurement (MPM). Journal of Quality in Maintenance Engineering. 2007;**13**(3):241-258. DOI: 10.1108/13552510710780276

[5] Parida A, Kumar U. Maintenance performance measurement (MPM): Issues and challenges. Journal of Quality in Maintenance Engineering. 2006; **12**(3):239-251. DOI: 10.1108/135525 10610685084 6

[6] Kans M, Inglwad A. Common database for cost-effective improvement of maintenance performance. International Journal of Production Economics. 2008;**113**(2):734-747. DOI: 10.1016/j.ijpe.2007. 10.008 7

[7] Sari E, Shaharoun AM, Ma'aram A, Yazid AM. Sustainable maintenance performance measures: A pilot survey in Malaysian automotive companies. Procedia CIRP. 2015;**26**:443-448. DOI: 10.1016/j.procir.2014.07.163

[8] Maletič D, Maletič M, Al-Najjar B, Gomiščček B. The role of maintenance in improving company's competitiveness and profitability: A case study in a textile company. Journal of Manufacturing Technology Management. 2014;**25**(4): 441-456. DOI: 10.1108/JMTM-04-2013-0033

[9] Rault R, Trentesaux D. Artificial intelligence, autonomous systems and robotics: Legal innovations. In: Borangiu T, Trentesaux D, Thomas A, Cardin O, editors. Service Orientation in Holonic and Multi-Agent Manufacturing. Cham: Springer; 2018. pp. 1-9. DOI: 10.1007/978-3-319-73751-5_1

[10] Leukel J, González J, Riekert M. Adoption of machine learning technology for failure prediction in industrial maintenance: A systematic review. Journal of Manufacturing Systems. 2021;**61**:87-96

[11] Shcherbakov MV, Glotov AV, Cheremisinov SV. Proactive and predictive maintenance of cyber-physical systems. In: Kravets A, Bolshakov A, Shcherbakov M, editors. CyberPhysical Systems: Advances in Design & Modelling. Vol. 259. Cham: Springer; 2020. pp. 263-278. DOI: 10.1007/978-3-030-32579-4_21

[12] Chaudhuri A. Predictive maintenance for industrial IoT of vehicle fleets using hierarchical modified fuzzy support vector machine. arXiv. 2018. DOI: 10.48550/arXiv.1806.09612

[13] Kamariotis A, Tatsis K, Chatzi E, Goebel K, Straub D. A metric for assessing and optimizing data-driven prognostic algorithms for predictive maintenance. Reliability Engineering & System Safety. 2024;**242**:109723, ISSN 0951-8320. DOI: 10.1016/j.ress.2023.109723

[14] Khazaelpour P, Zolfani SH. FUCOM-optimization based predictive maintenance strategy using expert elicitation and artificial neural network. Expert Systems with Applications. 2024; **238**(Part A):121322, ISSN 0957-4174. DOI: 10.1016/j.eswa.2023.121322

[15] Sherien E, Ahmed M, Abdelrahman S, et al. Machine learning and IoT–based predictive maintenance approach for industrial applications. Alexandria Engineering Journal. 2024; **88**:298-309

[16] Bencheikh G, Letouzey A, Desforges X. An approach for joint scheduling of production and predictive maintenance activities. Journal of Manufacturing Systems. 2022;**64**: 546-560

[17] Zonta T, da Costa CA, Zeiser FA, et al. A predictive maintenance model for optimizing production schedule using deep neural networks. Journal of Manufacturing Systems. 2022;**62**: 450-462

[18] Ruiz-Sarmiento JR, Monroy J, Moreno FA, Galindo C, Bonelo JM, Gonzalez Jimenez J. A predictive model for the maintenance of industrial machinery in the context of industry 4.0. Engineering Applications of Artificial Intelligence. 2020;**87**:103289. DOI: 10.1016/j.engappai.2019.103289

[19] Chakroun A, Hani Y, Masmoudi F, El Mhamedi A. Modèle prédictif pour l'évaluation de la santé d'une unité d'assemblage basé sur l'apprentissage automatique dans le contexte de l'industrie 4.0. In: 1 er Congrès de la Société Française d'Automatique, Génie Industriel et de Production SAGIP 2023, 7–9 Juin 2023, Marseille, France. 2023

[20] Chakroun A, Hani Y, Elmhamedi A, et al. A predictive maintenance model for health assessment of an assembly robot based on machine learning in the context of smart plant. Journal of Intelligent Manufacturing. 2024. DOI: 10.1007/s10845-023-02281-3

[21] Dopico M, Gómez A, De la Fuente D, García N, Rosillo R, Puche J. A vision of industry 4.0 from an artificial intelligence point of view. In: Proceedings on the International Conference on Artificial Intelligence (ICAI). The Steering Committee of the World Congress in Computer Science, Computer Engineering and Applied Computing (WorldComp); 2016. p. 407

Chapter 6

Navigating Sustainability: A Conceptual Exploration of Road Freight Decarbonisation

Anthony E. Ezenwa, Akunna Oledinma, Martin Osei, Obiajulu Ede, Christopher Ikeogu and Nze Ibeawuchi

Abstract

This chapter underscores the critical importance of embracing alternative fuels, specifically natural gas and hydrogen, as pivotal elements in the concerted effort to decarbonise road freight operations. It accentuates the necessity for collaborative, well-informed decision-making processes to navigate the complexities of this transition successfully. The exploration extends beyond technological considerations to encompass policy-driven initiatives adopted by nations like Norway, India, the USA, the UK, South Africa, and Germany, among others. Notably, it scrutinises instances of demonstrated economic viability associated with specific decarbonisation strategies employed in these regions. Moreover, the chapter advocates for sustained interdisciplinary research efforts to comprehensively address the diverse challenges posed by regional variations, industry-specific requirements, operational environments, cross-cultural influences, and the integration of sustainability metrics. By emphasising the importance of ongoing technological innovation, robust policy frameworks, and thoughtful economic considerations, the chapter envisions a global roadmap for sustainable road freight practices. In essence, it calls for a holistic approach, urging scholars, policymakers, and industry experts to collaboratively shape a sustainable future for road freight that addresses the intricate interplay of technological advancements, policy dynamics, and economic imperatives on a global scale.

Keywords: road freight decarbonisation, alternative fuels, natural gas, hydrogen, policy frameworks, sustainability metrics

1. Introduction

The urgent need to address climate change has led to a detailed examination of various industries' environmental impacts, highlighting transportation as a major contributor to greenhouse gas emissions [1–3]. Among transportation sectors, road freight operations stand out as significant contributors to carbon emissions [4, 5]. In India, road transportation alone accounted for more than 56% of the country's transportation sector emissions in 2019, totalling 2508 million tonnes (Mton), surpassing the global contribution of the entire transportation sector [2]. The transport sector in

IntechOpen

India represents 18% of total energy consumption, requiring an estimated 200 Mtoe of energy stock by 2030 if current trends persist [2].

Similarly, China, with the world's largest truck industry, sees over 50% of its total transport emissions originating from road freight [3]. Globally, the transport sector, representing 28% of total energy demand, is projected to contribute up to 10.3 Gt CO_2/year by 2040 [6]. Decarbonising this sector poses formidable challenges due to factors such as high vehicle dependency, widespread use of combustion-based engine technologies, low petroleum costs, and limited alternatives to fossil fuels, and fragmentated operation [1].

The Intergovernmental Panel on Climate Change [7] emphasises that the primary source of carbon emissions in road freight is the combustion of fossil fuels, particularly diesel, leading to the release of carbon dioxide (CO_2) and contributing to global climate change [4, 6, 8–10]. Secondary pollutants, such as nitrogen oxides (NOx) and particulate matter, further impact air quality and public health [2, 5, 11, 12]. Diesel-powered trucks operating in urban areas worsen air quality, posing a threat to public health [6].

Recognising the urgency of addressing this issue, Lee and Kim [13] emphasises the need to transform road freight operations through sustainable practices and technological innovations. Therefore, this chapter aims to provide a foundation for a concise exploration of road freight decarbonisation, covering challenges, strategic approaches, policy frameworks, technological innovations, case studies, and global perspectives. It aligns with the perspective that road freight's heavy reliance on fossil fuels necessitates global efforts to mitigate climate change and enhance environmental sustainability.

Subsequent sections of the chapter encompass methodology, a brief literature review, an investigation into challenges, approaches, and strategies to decarbonise road freight operations. Others comprise global perspectives, case studies showcasing best practices and lessons learned, an examination of future outlook, discussion, and ultimately, a conclusion providing guidance for future studies.

2. Methodology

This chapter employs a conceptual literature review method to furnish an exhaustive comprehension of the intricate domain of road freight decarbonisation [14]. Through amalgamating existing knowledge, this review seeks to elucidate fundamental concepts, burgeoning trends, and crucial lacunae in the literature, thereby establishing the groundwork for a nuanced exploration of sustainability in the realm of road freight. Specifically, the chapter establishes a lucid comprehension of sustainability within the ambit of road freight by scrutinising seminal works and contemporary literature on sustainability. It focuses on how sustainability is conceptualised and applied in the road freight industry. Subsequently, the approach identifies and scrutinises various strategies employed for decarbonising road freight, drawing upon empirical studies, industry reports, and technological advancements. These decarbonisation approaches are categorised into technological, operational, and policy-driven interventions.

Further, the chapter delves into the economic dimension of transitioning towards sustainable road freight practices. It draws on empirical papers, business case studies, and industry reports to fathom the financial implications for stakeholders in the road freight sector. Following this, the chapter evaluates the environmental consequences of road freight decarbonisation efforts through an examination of life cycle assessments, carbon footprint analyses, and ecological impact studies. This enables an assessment of the overall environmental sustainability of different decarbonisation strategies. Moreover, the chapter scrutinises the role of policies and regulations in shaping road freight decarbonisation by reviewing national and international policy documents. It analyses their effectiveness, challenges, and potential for promoting sustainable practices in the road freight industry. In a similar vein, the chapter identifies and discusses key challenges and associated opportunities with road freight decarbonisation. This is achieved through the synthesis of empirical findings to provide insights into the multifaceted nature of challenges and opportunities in this context.

Finally, the chapter incorporates diverse stakeholders' perspectives to enhance the conceptual understanding of road freight decarbonisation. This is accomplished through the exploration of industry reports and business case studies. In conclusion, the conceptual literature review culminates in a comprehensive synthesis, offering a holistic understanding of road freight decarbonisation within the broader context of sustainability. By traversing the conceptual landscape, this chapter aims to provide a robust foundation for subsequent discussions and analyses in the pursuit of a sustainable future for road freight.

3. Brief literature review

In the pursuit of sustainable road freight, recent studies (e.g., [1, 6, 8, 9, 15]) offer valuable insights into the multifaceted challenges and opportunities within this domain. The collective findings underscore the complexity of road freight decarbonisation, emphasising the intricate interplay of technological, economic, and policy dimensions. Churchman et al. [1] recognise the pivotal role of freight demand segments in influencing decarbonisation strategies. Varied preferences for motive technologies and the potential of containerised rail highlight the need for nuanced, context-specific approaches. Hossain et al. [2] delve into regional specifics, focusing on India's roadmap for 2050. Their emphasis on adopting alternative fuels and advanced technologies aligns with a global push towards sustainability.

Ma et al. [3] conducts a comprehensive Total Cost of Ownership (TCO) analysis, shedding light on the life cycle costs of electric and fuel cell (FC) trucks. The study's emphasis on emission reductions and the role of alternative fuels resonates with Gómez Vilchez et al. [4], who evaluate European decarbonisation trends, highlighting the dominance of natural gas and the imperative for unified policies. de las Nieves Camacho et al. [6] examine hydrogen heavy-duty FC trucks, stressing collaboration, and objective testing for informed decision-making. Haugen et al. [8] tackle hydrogen production challenges, recommending electrolysis as a short-term solution. They identify critical research domains, aligning with Zhang et al. [16], who optimise FC systems in commercial trucks. The potential of hydrogen extends to long-haul strategies, as illustrated by Rohith et al. [17], promoting the platooning of electrified heavy goods vehicles. Rosenberg et al. [18] examine Norway's policy-driven approach, while de Saxe et al. [15] provide evidence of tangible benefits from a 2020 project, demonstrating the economic viability of certain strategies. Barton and Thomson [10] contribute valuable insights into electrification feasibility based on vehicle types and energy needs. Deshpande et al. [19] propose Electric Road Systems (ERS) as economically viable for key transport networks. Studies by Zhang et al. [5] and Melo [9] emphasise the significance of regional and sector-specific approaches. Zhang et al. [5]

advocate for a balanced policy focus in Ireland, while Melo [9] empirical model in Scotland underscores the importance of demand- and supply-led factors in achieving emission reduction targets.

Capros et al. [20] advocate for a model-based analysis to decarbonise the EU economy, stressing the cost significance of timely access to decarbonisation options. Sharmina et al. [21] introduce a new framework for monitoring and assessing mitigation progress, aligning with the call for integrated models and highlighting untapped mitigation opportunities. In conclusion, this critical synthesis calls for continued interdisciplinary research, integrating technological innovation, regional specifics, policy frameworks, and economic considerations. Practical implementations and pilot projects, similar to de Saxe et al. [15], should be encouraged to validate theoretical models. Furthermore, a focus on social science aspects, as highlighted by Churchman and Longhurst [22] is crucial for understanding business transitions and ensuring consumer-centric sustainability. Future studies should aim for a holistic understanding of road freight decarbonisation, fostering collaboration between academia, industry, and policymakers to drive sustainable change.

Moreover, we explore current trends in academic publications within the road freight decarbonisation debate. This section focuses on presenting and summarising observed patterns, frequencies, and themes with limited subjective opinions. Our review of the timeline and article distribution indicates an increasing trend in research output, highlighting the subject's growing relevance. Most publications (9) are from Elsevier Ltd., followed by MDP and Taylor and Francis Ltd., with 4 and 2 publications, respectively, alongside others. Similarly, the review papers are 16 articles and 2 review papers in the specified context. The distribution of articles by journals provides an overview of the sources contributing to research in energy, transportation, and related environmental topics (see **Figure 1a**). The methodology employed in the reviewed articles encompass scholarly literature examination, quantitative analysis through engineering and statistical modelling, as well as simulation techniques, and qualitative investigation through interviews (see **Figure 1b**). Finally refer to **Table 1** for the full meaning of abbreviations.

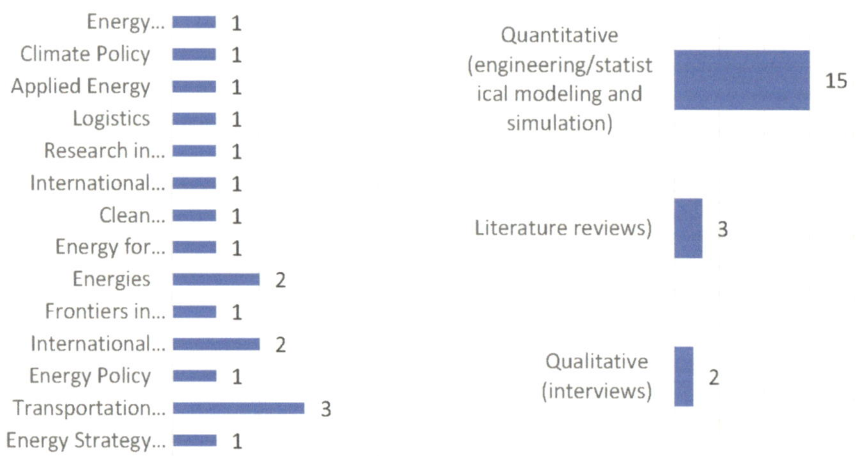

Figure 1.
Publication trend by journals and research methods.

Abbreviation	Full meaning
EV	Electric vehicles
IPPC	Intergovernmental Panel on Climate Change
IEA	International Energy Agency
WHO	World Health Organisation
ISO	International Organisation for Standardisation
ITF	International Trust Fund
USDE	U.S. Department of Energy
LNG	Liquefied natural gas
LRC	Low Carbon Logistics (LRC)
HGV	Heavy goods vehicles
LGV	Light goods vehicles
FCEV	Fuel cell electric vehicle
FC	Fuel cell
CO_2	Carbon dioxide
NOx	Nitrogen oxides
GHG	Greenhouse gas
ICEs	Internal combustion engines
CE	Circular economy
TCO	Total Cost of Ownership

Table 1.
Abbreviations and their full meanings with references.

4. Understanding the challenge

4.1 Emission sources in road freight

Understanding and mitigating road freight emissions necessitates addressing various sources across the supply chain, including vehicle fuel combustion, manufacturing, maintenance, infrastructure, idling inefficiencies, and suboptimal logistics practices [7, 13, 21, 22]. To formulate effective strategies, it is crucial to acknowledge and target these points. In road freight operations, where diesel in heavy-duty trucks is a major emission source [23], efficient fuel combustion is paramount (see **Figure 2**—Interlink transport coal in the PBS pilot project). This process, occurring within internal combustion engines, involves chemical reactions between fuel hydrocarbon molecules and oxygen, impacting carbon dioxide (CO_2) emissions per unit of energy [24]. Modern trucks use advanced technologies to enhance combustion efficiency and reduce emissions [25]. Life cycle assessment (LCA) methodologies, considering the entire life cycle, guide sustainable solutions [24]. Environmental impact regulations, like those by the European Commission, aim to mitigate traditional fuel combustion effects [26]. With a growing emphasis on alternative fuels like electric vehicles, hydrogen fuel cells, and biofuels [23], the road to decarbonising road freight involves considerations of fuel combustion efficiency, emissions, regulatory frameworks, alternative fuels, technological innovations, and holistic environmental impact assessments.

Figure 2.
Interlink transport coal in the PBS pilot project (source: [15]).

Vehicle manufacturing and maintenance are integral to road freight's carbon footprint. Energy-intensive production processes and material extraction contribute to emissions. Life cycle assessment (LCA) studies, such as Hawkins, Singh [24], provide insights into the environmental impact of manufacturing. Material selection and lightweighting strategies inform emission rates [25]. Efficient maintenance practices, including predictive maintenance, can reduce emissions [27] Circular economy principles in manufacturing and maintenance optimise designs for longevity and recycling, mitigating environmental impact [28]. Decarbonising road freight involves a shift to electric vehicles, with the [7] highlighting their manufacturing and maintenance efficiency. Optimising manufacturing processes and embracing sustainable maintenance practices contribute to a more environmentally sustainable road freight industry.

Infrastructure construction and maintenance, energy-intensive processes, contribute substantially to emissions [29]. Integrating intelligent technologies like condition monitoring and predictive maintenance is essential [30]. Sustainable practices, such as incorporating alternative materials and construction technologies, are pivotal [31]. Green corridors, wildlife-friendly structures, and strategic investments in multimodal transportation systems are essential for sustainable infrastructure [32]. Infrastructure development and maintenance need careful consideration for road freight decarbonisation. Implementing sustainable practices, exploring alternative materials, and adopting innovative technologies are fundamental steps towards a more ecologically sound road freight industry.

Vehicle idling and operational inefficiencies significantly contribute to carbon emissions in road freight. Reducing idling time, optimising routes, and improving load factors are direct approaches to mitigate emissions [33]. Technologies like telematics and connectivity enable real-time monitoring, contributing to decarbonisation by enhancing efficiency [34]. Smart logistics, incorporating real-time tracking and supply chain optimisation through IoT and AI, addresses inefficiencies [35]. Idling regulations, anti-idling technologies, and emission standards drive the adoption of fuel-efficient technologies [26, 36]. Addressing idling and operational inefficiencies is pivotal for road freight decarbonisation, and integrating technological solutions, smart logistics, and regulatory measures contributes to sustainable freight operations.

Inefficient cargo capacity utilisation and the transportation of empty containers contribute to increased carbon emissions. Optimising return journeys, implementing smart logistics, and collaborative initiatives enhance operational efficiency [37, 38]. Integrating technological solutions like telematics and route optimisation is essential. Regulatory measures incentivising or mandating reduced empty haulage and underutilised capacity contribute to road freight decarbonisation [38]. A multifaceted approach, including technological solutions, collaborative freight initiatives, and regulatory measures, is essential for mitigating these inefficiencies and achieving road freight decarbonisation.

Inefficient logistics practices, including suboptimal route planning and excessive inventory, contribute to increased carbon emissions. Efficient resource utilisation, technology integration, alternative fuel adoption, green logistics initiatives, collaboration, and regulatory compliance are crucial elements driving sustainable practices in road freight [13]. Aligning logistics and supply chain practices with environmental goals is a critical prerequisite for achieving meaningful decarbonisation. Likewise, a comprehensive understanding of diverse emission sources is essential for designing effective decarbonisation strategies for road freight.

4.2 Lifecycle assessment (LCA) in road freight

Life cycle assessment (LCA) is a holistic approach that evaluates the environmental impact of road freight activities from production to end-of-life, covering stages such as raw material extraction, manufacturing, transportation, use, and disposal [39]. LCA identifies key points in the life cycle, directing decarbonisation efforts towards impactful stages like fuel combustion, manufacturing, and logistics [39]. Additionally, LCA facilitates the comparison of technologies, fuels, and operational practices, aiding decisions on cleaner alternatives in the road freight sector [39].

LCA's significance lies in exposing carbon emissions associated with fuel combustion in road freight vehicles. Decarbonisation guided by LCA can focus on enhancing combustion efficiency, transitioning to alternative fuels, and adopting electric or hybrid vehicles [24]. Manufacturing emerges as a vital contributor to life cycle emissions, and LCA guides the selection of sustainable materials, lightweighting strategies, and eco-friendly manufacturing processes to reduce environmental impact [24]. LCA underscores logistics and supply chain practices as critical components affecting overall environmental impact. Insights can drive the adoption of smart logistics strategies, route optimisation, and collaborative efforts within the supply chain to enhance decarbonisation [40].

Considering the entire life cycle, LCA supports the argument for extending the operational lifespan of road freight vehicles. Maintenance and repair strategies, informed by LCA, contribute to reducing the need for frequent replacements and associated manufacturing emissions [41]. In summary, life cycle assessment stands as a crucial tool for evaluating and comprehensively addressing the environmental impact of road freight operations. Its holistic perspective, hotspot identification, and comparative analysis offer valuable insights crucial for effective decarbonisation, supporting optimisation of fuel combustion, sustainable material selection, efficient logistics practices, and the extension of vehicle lifespan—critical elements in achieving meaningful road freight decarbonisation.

5. Strategic approaches and global perspective

5.1 Strategic approaches

Decarbonising road freight is a complex challenge demanding strategic solutions across the entire supply chain. This assessment explores key strategies such as alternative fuels, advanced vehicles, modal shift, intermodal optimisation, green logistics initiatives, and regulatory compliance. Each strategy is evaluated for effectiveness, feasibility, and broader industry impact.

- *Alternative fuels*: the integration of alternative fuels, including liquefied natural gas (LNG), hydrogen, and electricity, stands out as a pivotal strategy for reducing carbon emissions in road freight. Despite challenges such as infrastructure needs and initial costs, alternative fuels significantly contribute to lowering the carbon intensity of freight operations. The success of this strategy is amplified when combined with logistics optimisation, advanced vehicle technologies, and regulatory compliance.

- *Advanced vehicle technologies*: innovations in engine efficiency, aerodynamics, and drivetrain optimisation play a crucial role in enhancing fuel efficiency and aligning with decarbonisation goals. Challenges like upfront costs and charging infrastructure gaps exist, but the overall impact on emission reduction is noteworthy. The synergy of advanced vehicle technologies with alternative fuels and logistics optimisation maximises their effectiveness in addressing various facets of carbon emissions in the road freight sector.

- *Modal shift and intermodal optimisation*: modal shift involves redirecting freight to more sustainable modes, such as rail and water transport. Intermodal optimisation integrates different transportation modes for efficiency gains. Despite challenges like infrastructure limitations, both strategies complement each other, offering a diversified and environmentally sustainable freight transportation system. Successful implementation requires supportive policies, regulations, and strategic planning.

- *Green logistics initiatives*: green logistics efforts focus on optimising vehicle fleets, adopting alternative fuels, and leveraging intermodal transport. Energy-efficient technologies and last-mile delivery solutions contribute to decreasing emissions per unit of freight transported [42, 43]. Digital technologies, including real-time tracking and data analytics, play a pivotal role in enhancing efficiency. Overcoming challenges requires collaborative efforts from stakeholders, governments, and policymakers.

- *Regulatory compliance and advocacy*: regulatory compliance involves meeting emission standards to ensure environmentally acceptable road freight operations. Advocacy complements this by influencing policymakers to institute incentives and support the transition to cleaner alternatives. Effective advocacy addresses challenges related to enforcement and industry resistance. International collaboration is essential for harmonising regulations globally and achieving meaningful reductions in carbon emissions.

In conclusion, decarbonising road freight requires a holistic and integrated approach that combines these strategies. Each element contributes uniquely, and their collective implementation is key to achieving substantial and sustainable reductions in carbon emissions within the global road freight sector. Policymakers, industry stakeholders, and international bodies must collaborate to create an environment conducive to embracing these decarbonisation measures.

5.2 Global perspectives

This section explores the pivotal role of decarbonising road freight in addressing global climate change and sustainability. The literature review sheds light on key trends, challenges, and opportunities influencing the international landscape. A notable global trend is the increasing adoption of electric vehicles (EVs) in road freight, backed by significant investments from governments and industries, essential for reducing reliance on fossil fuels and mitigating carbon emissions. Additionally, the integration of advanced technologies like autonomous and connected vehicles, telematics, and real-time analytics is gaining traction globally, enhancing operational efficiency and fuel savings.

However, a significant challenge on a global scale is the insufficient charging infrastructure for EVs, emphasising the need for a robust and accessible charging network to fully exploit the potential of electric road freight [36]. Inconsistencies in policies across regions present obstacles to global road freight decarbonisation, underscoring the importance of harmonising regulations and emission standards worldwide [26]. Due to the global nature of road freight, increased international collaboration is considered necessary, enabling the sharing of best practices, technological innovations, and policy frameworks to expedite the decarbonisation process [44].

Opportunities lie in promoting multimodal transportation solutions integrating road, rail, and waterway transport, with coordinated global efforts having the potential to optimise supply chain logistics and contribute to overall emission reduction [37]. Stringent emission standards are becoming prevalent worldwide, with governments recognising the necessity of regulating road freight vehicle emissions to achieve climate goals [26]. Many countries are introducing incentives and subsidies to encourage the adoption of cleaner technologies in road freight, aligning with the global push for sustainable transportation [45]. The global outlook for road freight decarbonisation reflects a dynamic landscape, marked by a shift towards electric

Region	Challenges	Opportunities	Strategies
North America	• Lack of comprehensive EV charging infrastructure • Policy fragmentation	• Abundance of renewable resources • Innovation hubs	• Collaboration on charging infrastructure • Advocacy for unified emission standards
Europe	• Urban congestion • Transition period for traditional fleets	• Extensive rail and waterway networks • Strong regulatory framework	• Urban logistics innovations • Incentive programs
Asia-Pacific	• High urbanisation rates • Diverse economic realities	• Growing middle class • Innovation in emerging economies	• Sustainable urban planning • Technology transfer and collaboration
Africa	• Insufficient road infrastructure • Economic constraints	• Leapfrogging technologies • Infrastructural development	• Regional collaboration • Economic empowerment

Table 2.
Overview of the regional road freight decarbonisation efforts.

vehicles, advanced technology integration, and supportive policies. Despite challenges, international collaboration and concerted efforts can unlock opportunities for sustainable road freight operations worldwide. In summary, this section underscores the global necessity of decarbonising road freight while highlighting variations in challenges, opportunities, and strategies across distinct regions, focusing on North America, Europe, Asia-Pacific, and Africa (refer to **Table 2**).

6. Case studies of best practices, lessons learned, and future outlook

6.1 Case studies of best practices and policies

HJF Transports BVBA, a burgeoning family enterprise in Belgium with around thirty trucks, grapples with operational cost management and carbon emission reduction challenges [43]. In a strategic alliance with Shell Belgium, a prominent natural gas supplier, the company is transitioning its truck fleet to liquefied natural gas (LNG), perceived as a competitive and sustainable alternative to diesel. As per Shell's report, 11 out of HJF Transports' 30 trucks exclusively run on LNG, with plans for further investment in the LNG network across Belgium, the Netherlands, Germany, France, and Poland. Mr. Fossoul, the company's representative, underscores the limited public awareness of LNG's efficiency and sustainability, emphasising the imperative for increased investment in this alternative [46].

Amid uncertainties and mounting diesel costs, HJF Transports perceives LNG as the optimal solution for immediate and sustainable fleet operations. The company's commitment to environmental sustainability aligns with its 2018 investment strategy in LNG-powered trucks, a decision made in partnership with Shell Belgium. The success of the current LNG-powered trucks validates the company's choice to invest in a more sustainable solution. The ongoing collaboration between HJF Transports and Shell is characterised as a true partnership, with Shell incorporating customer requests and operational realities into the development of new services. HJF Transports plans to expand its LNG-powered fleet, leveraging Shell's growing pan-European LNG network facilitated by the Shell LNG card [46]. The company's proactive approach demonstrates its dedication to a sustainable business model amidst the challenges of rising diesel costs and environmental responsibility (see **Figure 3**).

Likewise, according to the Foresignt [47], Low Carbon Logistics (LRC) research has significantly influenced UK government policies, transforming the official approach to carbon auditing road freight transport [47]. It has offered guidance to the industry on measuring and managing CO_2 emissions, establishing the industry and government-endorsed target to reduce the carbon intensity of road freight transport. Responding to this research, the Freight Transport Association (FTA) initiated the Logistics Carbon Reduction Scheme (LCRS) in 2010. Collaborating with the FTA, the LRC adapted its freight decarbonisation model to meet LCRS needs and assessed how the UK road freight system should transform to achieve the CO_2 reduction target set by the 2008 Climate Change Act.

The research's impact is evident in various arenas, including influencing the Department of Energy and Climate Change (DECC)'s 2050 Pathways Analysis, which relied on LRC's work for setting emission levels. The LRC's freight decarbonisation model played a pivotal role in shaping the Fourth Carbon Budget by the Committee on Climate Change (CCC), contributing to the UK's efforts to tackle climate change. Additionally, the LRC's analytical framework for assessing logistics-related

Figure 3.
HJT vehicle fleet and Shell LNG plant (source: [46]).

externalities at a country level has been adopted and refined by large EU-funded projects and applied in reports for the UK Commission for Integrated Transport. Furthermore, the LRC's research scrutinised CO_2 emissions from road haulage operations, prompting methodological changes in compiling CO_2 data for the National Atmospheric Emissions Inventory. The research demonstrated that decarbonising the road freight sector might be less challenging than anticipated. Notably, the LRC's work has influenced industry practices, government policies, and international initiatives, showcasing its multifaceted and profound impact on the road freight sector's journey towards decarbonisation.

6.2 Lessons learned

The road freight sector, a linchpin in global trade, faces a transformative shift towards sustainability due to environmental concerns like carbon emissions and air pollution. Global initiatives are underway to usher in ecologically conscious practices, encompassing technological integration, collaborative efforts, regulatory backing, and innovative financial models. Workforce training, gender equality, lifecycle assessment, sustainability metrics, and adaptability are also crucial dimensions. Electric vehicles (EVs) are key to reducing carbon emissions. Azapagic and Perdan [48] emphasise electric trucks and vans powered by batteries or hydrogen fuel cells as alternatives to diesel. Hybrid systems, combining internal combustion engines with electric power, enhance fuel efficiency. Fuel-efficient technologies, aerodynamic enhancements, lightweight materials, and advanced engine management systems optimise fuel consumption. Telematics and connectivity solutions use real-time data analytics for efficient fleet management. Autonomous vehicles offer increased efficiency and reduced fuel consumption.

Exploring alternative fuels like compressed and liquefied natural gas provides cleaner-burning options. In urban logistics, electric cargo bikes and drones reduce congestion and emissions. Successful decarbonisation relies on strategic technology selection, adaptability, investment planning, financial viability, data-driven decision-making, stakeholder collaboration, employee training, change management, regulatory compliance, scalability, replicability, continuous improvement, and innovation. Selecting technologies based on operational needs, adapting to innovations, and conducting thorough assessments contribute to successful integration. Infrastructure planning and collaboration are crucial for an accessible charging network. Real-time data analytics and predictive modelling optimise routes and reduce energy consumption. Investing in training programs, robust financial planning, and supportive regulations are key factors for success. Collaboration with stakeholders, continuous monitoring of technology performance, flexibility, life cycle assessments, and transparent sustainability metrics enhance stakeholder trust. Pilot programs assessing feasibility and effectiveness contribute to widespread impact. Lessons from successful initiatives underscore the importance of these factors in the road freight sector's journey towards decarbonisation.

6.3 Future outlook

The future of road freight decarbonisation holds promise with the integration of emerging technologies and anticipated policy developments. This outlook envisions a transformative shift towards sustainable practices, driven by advancements in technology and a supportive policy landscape.

6.3.1 Emerging technologies

The electric vehicle (EV) landscape is set for rapid growth, fuelled by advancements in batteries, improved charging infrastructure, and increased energy density (see **Figure 4**). Solid-state batteries and fast-charging technologies, highlighted by Rai et al. [49], promise to enhance the viability of EVs in road freight. Meanwhile, the integration of autonomous and connected technologies, as discussed by Thomopoulos et al. [50], is poised to revolutionise route planning and fleet efficiency. Ongoing research into alternative fuels like green hydrogen and advanced biofuels, along with innovations in propulsion systems, offers credible substitutes for traditional diesel. The potential of hydrogen fuel cells and improved engine designs plays a pivotal role in achieving carbon-neutral road freight, according to Kühnbach et al. [51].

6.3.2 Anticipated policy development

The future policy landscape is expected to introduce stricter global emission standards for road freight vehicles, fostering the adoption of cleaner technologies and penalising carbon-intensive practices [26]. Governments are likely to provide increased incentives and subsidies for the transition to low-emission and zero-emission vehicles, particularly in areas like fleet electrification, charging infrastructure

Figure 4.
An overhead centenary implementation of ERS. Source: Siemens Mobility GmbH, 2022, as cited in [19].

development, and sustainable transportation research [23]. Anticipated policies also suggest a shift towards integrated multimodal transportation solutions, promoting a balanced combination of road, rail, and waterway transport to reduce the overall carbon footprint [38]. Looking ahead, a collaborative approach leveraging emerging technologies and evolving policies, especially the convergence of electric and autonomous vehicles, is poised to drive innovation in the road freight sector, contributing significantly to global decarbonisation efforts.

7. Discussion and conclusion

7.1 Discussion

The chapter presents a stimulating analysis of the intricate aspects surrounding the sustainability of road freight. Employing a conceptual review method, it sheds light on key concepts, emerging trends, and critical gaps, delivering a nuanced exploration of sustainability concerns within the road freight sector. Commencing with the establishment of a clear understanding of sustainability within the context of road freight, the chapter engages with seminal works and contemporary literature to delve into the conceptualisation and application of sustainability. This not only lays the foundation for subsequent discussions but also emphasises the significance of defining sustainability in the evolving landscape of freight transportation.

A pivotal facet of the chapter involves the identification and analysis of diverse strategies for decarbonising road freight, categorised into technological, operational, and policy-driven interventions [25, 37, 43]. The implications extend beyond immediate environmental concerns, encompassing the socio-economic dimension of the freight sector and stressing the necessity for a multifaceted approach in addressing challenges posed by carbon emissions. The exploration of the economic dimension enriches the discourse by spotlighting the financial implications for stakeholders in the road freight sector [28]. This economic scrutiny aligns with the broader discourse on the green economy, highlighting the imperative of sustainable practices that strike a balance between environmental and economic considerations [37].

Moreover, the chapter critically assesses the environmental consequences of road freight decarbonisation efforts, utilising lifecycle assessment, carbon footprint analyses, and ecological impact studies [24, 40, 41, 48]. The implications extend beyond immediate environmental impact, contributing to the broader discourse on sustainable practices in transportation and emphasising the interconnectedness of environmental concerns with global sustainability goals. By scrutinising the role of policies and regulations, the chapter underscores the significance of government intervention in shaping road freight decarbonisation [21, 44]. This aligns with broader literature discussing the regulatory framework's role in promoting sustainable practices across industries, emphasising the need for collaborative efforts between policymakers and industry stakeholders. The identification and discussion of challenges and opportunities associated with road freight decarbonisation add a layer of realism to the conceptual exploration [26, 37, 42, 45]. This discussion aligns with broader literature on sustainability marked by both obstacles and potential breakthroughs.

Furthermore, the incorporation of diverse stakeholder perspectives enriches the conceptual understanding of road freight decarbonisation [31, 33, 35, 46, 47]. This emphasis on stakeholder engagement resonates with broader discussions in the literature on the importance of inclusive decision-making processes for effective sustainability initiatives. In summary, the implications of the issues covered in this chapter resonate with broader literature on sustainable transportation, green logistics, and the global shift towards more eco-friendly practices [37]. The interconnectedness of environmental, economic, and social aspects aligns with the holistic approach advocated in the broader literature [33, 37]. Against this backdrop, the implications highlighted underscore the urgency of addressing road freight sustainability as an integral part of broader global sustainability efforts.

7.2 Conclusion and direction for future studies

In the quest for a sustainable future in road freight, this chapter underscores the complex nature of decarbonisation, emphasising the need to tailor strategies to specific regional and industry contexts while integrating innovation and community engagement. The findings advocate for comprehensive approaches, highlighting the importance of clear reporting, sustainability metrics, and cross-cultural understanding to facilitate effective road freight adaptations. The review serves as a guiding resource, steering stakeholders towards collaborative solutions and stressing the necessity to customise decarbonisation efforts across diverse regions, industries, and operational contexts. Industry-specific approaches, aligned with operational environments, are deemed crucial for ensuring supply chain resilience. The roles of technological integration, cultural awareness, community engagement, and sustainability metrics are pivotal in optimising the environmental impact of road freight [37].

The future study agenda recommends prioritising tailored strategies for different regions, industries, and operational contexts to enhance the efficacy of decarbonisation initiatives, streamline logistics, and address specific challenges across diverse settings. The comprehensive research dimensions include investigating regional variations, analysing industry-specific adaptations, assessing operational environment considerations, scrutinising cross-cultural perspectives, and evaluating the integration of sustainability metrics using mainly qualitative research approach to help map the 'how' issues affecting road freight decarbonisation. Additionally, the chapter calls for future research to encompass emissions quantification, lessons learned transferability, sustainability beyond carbon reduction, ongoing commitment to learning, obstacle navigation, benchmarking, inspiration for similar efforts, and examining advocacy resulting from successful initiatives. In summary, adopting a multidimensional research approach is crucial for comprehensively addressing regional, industry-specific, and operational variations, contributing to the development of targeted and context-specific strategies for sustainable road freight practices globally.

Acknowledgements

Sincere appreciation to Dr. Anthony Whiteing, Professor Edward Sweeney, and Professor Jeffery Bray for their invaluable guidance and mentorship during the formulation of this chapter. The authors acknowledge the use of Chat GPT-3.5 for the language polishing of the manuscript.

Conflict of interest

None.

Author details

Anthony E. Ezenwa[1*], Akunna Oledinma[2], Martin Osei[1], Obiajulu Ede[1], Christopher Ikeogu[3] and Nze Ibeawuchi[3]

1 Bournemouth University, Bournemouth, United Kingdom

2 University of Hull, Hull, United Kingdom

3 Federal University of Technology Owerri, Nigeria

*Address all correspondence to: anthonyezenwa@ymail.com

References

[1] Churchman P, Dekker T, Pangbourne K. Transition codesign for purposive road freight decarbonisation. Transportation Research Part D: Transport and Environment. 2023;**125**:103980

[2] Hossain M et al. Narrowing fossil fuel consumption in the Indian road transport sector towards reaching carbon neutrality. Energy Policy. 2023;**172**:113330

[3] Ma X et al. Application of fuel cell and alternative fuel for the decarbonisation of China's road freight sector towards carbon neutral. International Journal of Hydrogen Energy. 2024;**49**:263-275

[4] Gomez VJ et al. An Analysis of Trends and Policies Supporting Alternative Fuels for Road Freight Transport in Europe. Frontiers in Energy Research. 2022;**10**:897916

[5] Zhang X, Plant E, Valantasis Kanellos N. An evaluation of ireland's sustainable freight transport policy. Logistics. 2022;**6**(3):65

[6] de las Nieves Camacho M, Jurburg D, Tanco M. Hydrogen fuel cell heavy-duty trucks: Review of main research topics. International Journal of Hydrogen Energy. 2022;**47**(68):29505-29525

[7] Masson-Delmotte V et al. Global Warming of 1.5 C: IPCC special report on impacts of global warming of 1.5 C above pre-industrial levels in context of strengthening response to climate change, sustainable development, and efforts to eradicate poverty. Cambridge University Press; 2022

[8] Haugen MJ et al. Electrification versus hydrogen for UK road freight: Conclusions from a systems analysis of transport energy transitions. Energy for Sustainable Development. 2022;**68**:203-210

[9] Melo PC. Driving down road transport CO_2 emissions in Scotland. International Journal of Sustainable Transportation. 2016;**10**(10):906-916

[10] Barton JP, Thomson M. Solar power and energy storage for decarbonisation of land transport in India. Energies. 2021;**14**(24):8277

[11] Munoz MN, Ballantyne EE, Stone DA. Assessing the economic impact of introducing localised PV solar energy generation and energy storage for fleet electrification. Energies. 2023;**16**(8):3570

[12] Pardhi S et al. A review of fuel cell powertrains for long-haul heavy-duty vehicles: Technology, hydrogen, energy and thermal management solutions. Energies. 2022;**15**(24):9557

[13] Lee S, Kim BG. Attribute of big data analytics quality affecting business performance. Journal of Social Computing. 2023;**4**(4):357-381

[14] Onu D, Kessler T, Smith JR. Admiration: A conceptual review. Emotion Review. 2016;**8**(3):218-230

[15] de Saxe C et al. High-capacity coal trucks to reduce costs and emissions at South Africa's power utility. Research in Transportation Business & Management. 2022;**48**:100898

[16] Zhang L et al. Design and optimisation of compressor for a fuel cell system on a commercial truck under real driving conditions. In: 15th International Conference on Turbochargers and Turbocharging. Howick Place, London: CRC Press; 2023

[17] Rohith G et al. Sustainable heavy goods vehicle electrification strategies for long-haul road freight transportation. IEEE Access. 2023;**11**:26459-26470

[18] Rosenberg E et al. Modelling the interaction between the energy system and road freight in Norway. Transportation Research Part D: Transport and Environment. 2023;**114**:103569

[19] Deshpande P et al. A breakeven cost analysis framework for electric road systems. Transportation Research Part D: Transport and Environment. 2023;**122**:103870

[20] Capros P et al. Model-based analysis of decarbonising the EU economy in the time horizon to 2050. Energy Strategy Reviews. 2012;**1**(2):76-84

[21] Sharmina M et al. Decarbonising the critical sectors of aviation, shipping, road freight and industry to limit warming to 1.5-2 C. Climate Policy. 2021;**21**(4):455-474

[22] Churchman P, Longhurst N. Where is our delivery? The political and socio-technical roadblocks to decarbonising United Kingdom road freight. Energy Research & Social Science. 2022;**83**:102330

[23] Anita WM et al. Interplay of climate change and air pollution-projection of the under-5 mortality attributable to ambient particulate matter (PM2. 5) in South Asia. Environmental Research. 2024;**21**:118292

[24] Hawkins TR et al. Comparative environmental life cycle assessment of conventional and electric vehicles. Journal of Industrial Ecology. 2013;**17**(1):53-64

[25] Askin AC et al. The heavy-duty vehicle future in the United States: A parametric analysis of technology and policy tradeoffs. Energy Policy. 2015;**81**:1-13

[26] Barsali S et al. Managing BEV charge to obtain a positive impact on a national power system. Energies. 2024;**17**(2):348

[27] Proost S, Van Dender K. What long-term road transport future? Trends and policy options. Review of Environmental Economics and Policy. 2011;**5**:44-65

[28] Yuan Z, Bi J, Moriguichi Y. The circular economy: A new development strategy in China. Journal of Industrial Ecology. 2006;**10**(1-2):4-8

[29] Merchan AL et al. Life cycle externalities versus external costs: The case of inland freight transport in Belgium. Transportation Research Part D: Transport and Environment. 2019;**67**:576-595

[30] Saravia de los Reyes R et al. Carbon footprint of a port infrastructure from a life cycle approach. International Journal of Environmental Research and Public Health. 2020;**17**(20):7414

[31] Mekong-Institute. Guidelines for Enhancing Freight Efficiency in ASEAN. 2015. Available from: https://www.mekonginstitute.org/what-we-do/ongoing-projects/greenfreightandlogisticsdevelopmentinmekongcountries/

[32] Allen J et al. The role of urban consolidation centres in sustainable freight transport. Transport Reviews. 2012;**32**(4):473-490

[33] US-Department-of-Energy. Idle Reduction for Heavy-Duty Trucks: A Way to Reduce Fuel Use and Emissions. 2016. Available from: https://www.energy.gov/eere/buildings/2016-results

[34] Pan S et al. The air quality and health impacts of projected long-haul truck and rail freight transportation in the United States in 2050. Environment International. 2019;**130**:104922

[35] McKinsey-Company. The Road to Sustainable Urban Logistics. 2019. Available from: https://www.mckinsey.com/about-us/new-at-mckinsey-blog/top-ten-insights-of-2019

[36] Allaoui H et al. Reducing transportation greenhouse gas emissions with collaborative distribution. Management Research Review. 2015;**38**(10):1049-1067

[37] McKinnon A et al. Green Logistics: Improving the Environmental Sustainability of Logistics. London: Kogan Page Publishers; 2015

[38] ITF. Towards Geater Resilience. 2015. Available from: https://www.worldbank.org/en/publication/trust-fund-annual-report-2021

[39] Seifert C et al. Life cycle assessment as decision support tool for environmental management in hospitals: A literature review. Health Care Management Review. 2021;**46**(1):12-24

[40] Axsen J, Hardman S, Jenn A. What do we know about zero-emission vehicle mandates? Environmental Science & Technology. 2022;**56**(12):7553-7563

[41] Li X et al. Optimal design of pure battery electric bus system on the grid network. Transportmetrica A: Transport Science. 2022;**20**:1-34

[42] Ezenwa A et al. Factors influencing information and communication technology diffusion in Nigeria's transport logistics industry: An exploratory study. International Journal of Integrated Supply Management. 2020;**13**(2-3):252-276

[43] Ezenwa A et al. Development of strategies to improve information communication technology diffusion in Nigeria's logistics and transport industry: Adaptation of structure-process-outcome model. International Journal of Integrated Supply Management. 2021;**14**(4):363-391

[44] Holguín-Veras J. Impacts of Freight Parking Policies in Urban Areas: The Case of New York City. New York: University Transportation Research Center - Region. 2016;**2**:40

[45] IEA, Global EV Outlook, I.E. Agency, editors. 2023. Available from: https://www.iea.org/topics/global-ev-outlook

[46] Shell-CRT. Shell CRT news, reports & case studies. Shell. 2023. Available from: https://www.shell.com/business-customers/shell-commercial-road-transport/news-and-casestudies.html

[47] Foresignt, Future of Mobility: Decarbonising road freight [Internet]. G.O.f. Science, editor. 2019. Available from: https://assets.publishing.service.gov.uk/media/5c6fe3b040f0b647af8f35e2/decarbonising_road_freight.pdf

[48] Azapagic A, Perdan S. Indicators of sustainable development for industry: A general framework. Process Safety and Environmental Protection. 2000;**78**(4):243-261

[49] Rai HB et al. Improving urban freight transport sustainability: Policy assessment framework and case study. Research in Transportation Economics. 2017;**64**:26-35

[50] Thomopoulos N et al. All work and no play? Autonomous vehicles and non-commuting journeys. Transport Reviews. 2021;**41**(4):456-477

[51] Kühnbach M et al. Impact of electric vehicles: Will German households pay less for electricity? Energy Strategy Reviews. 2020;**32**:100568

Chapter 7

Investigating How Logistics Operations Affect an Organization's Financial Performance

Dennis Sebata

Abstract

The fact that this area requires more attention is becoming more widely acknowledged. An organization's ability to operate efficiently depends heavily on logistics. A detailed examination of prospective observational studies indicates that successful financial control over transporting products and services from their point of origin to their final destination is essential for a business. In addition to offering guidance on how businesses can streamline their logistics processes to increase profitability, this study looks at how logistics functions affect financial performance. A company's financial success is significantly impacted by its logistics operations, ensuring that customers receive their orders on time is one way that efficient logistics management can raise revenue, minimize inventory costs, and save transportation expenses. To maximize the effectiveness of their logistics strategies, businesses need to examine every aspect of their supply chain from top to bottom. Using cutting-edge technologies, such as artificial intelligence and machine learning in the guidelines, means that businesses can use data analytics to ensure best practices, strengthen relationships with suppliers and partners, and gain insights into their supply chains' overall performance. However, this strategy necessitates the achievement of an optimal solution. These are the main conclusions of this work that are compiled and presented. Through a focus on streamlining operations and optimizing their logistics strategies, businesses can improve their financial performance.

Keywords: business functions, financial performance, logistics operations, management processes, measuring performance

1. Introduction

In the rapidly evolving business landscape, logistics has emerged as a critical component that ensures the smooth operation of enterprises [1]. Modern logistics operations are comprehensive procedures that involve the movement of goods and services from their point of origin to their final destination [2, 3]. This intricate management process encompasses vital tasks such as transportation, inventory management, warehousing, and packaging. Each of these tasks plays a significant role in determining an organization's financial performance [3]. Therefore, a detailed examination of the impact of logistics operations on financial performance can reveal potential opportunities for businesses to refine their logistics processes, thereby enhancing profitability [4].

The financial performance of a business is intrinsically linked to the efficiency of its logistics operations. Factors, such as customer satisfaction, production costs, and delivery schedules, are all influenced by the handling and movement of goods and materials [1, 3]. Efficient logistics management can lead to a reduction in inventory and transportation costs and an increase in revenue by ensuring the timely delivery of goods to customers. As such, it is imperative for businesses to invest in robust logistics strategies to enhance their supply chain efficiency and operational effectiveness, ultimately leading to higher financial returns [2].

In the fiercely competitive business environment of today, companies are perpetually on the lookout for strategies to increase their profits and gain a competitive edge [2, 5]. One such strategy is the simplification of their supply chain logistics. By streamlining their logistics processes, organizations can increase profitability, reduce costs, and improve overall efficiency [6].

To optimize their logistics operations, businesses must adopt a comprehensive strategy that involves a thorough analysis of every aspect of their supply chain [4, 6]. This includes understanding the flow of materials from suppliers to factories, managing inventory levels, coordinating with transportation providers for timely delivery, and ensuring the products reach the end customers in the best possible condition [3, 7]. By doing so, businesses can identify bottlenecks, inefficiencies, and areas of improvement in their logistics operations, leading to cost savings, improved customer satisfaction, and ultimately, enhanced financial performance.

In conclusion, logistics operations play a pivotal role in determining a business's financial performance. By investing in efficient logistics strategies and continuously refining their logistics processes, businesses can significantly enhance their profitability and gain a competitive edge in the market [3]. This chapter aims to delve deeper into this topic and provide a comprehensive understanding of the relationship between logistics operations and financial performance.

The importance of logistics in today's business world cannot be overstated. It is the backbone that supports the flow of goods and services from the point of origin to the point of consumption. It involves a multitude of tasks, each of which contributes to the overall efficiency and profitability of a business. These tasks include transportation, inventory management, warehousing, and packaging. By optimizing these tasks, businesses can significantly improve their financial performance.

The financial performance of a business is directly influenced by the efficiency of its logistics operations. A well-managed logistics system can lead to reduced inventory and transportation costs, increased revenue through timely delivery of goods to customers, and improved customer satisfaction [5, 6]. Therefore, it is crucial for businesses to invest in effective logistics strategies that can enhance their supply chain efficiency and operational effectiveness.

Today, in the highly competitive business environment, companies are constantly seeking ways to increase their profits and gain a competitive edge. One effective strategy is to simplify their supply chain logistics. By streamlining their logistics processes, companies can increase their profitability, reduce costs, and improve overall efficiency.

To optimize their logistics operations, businesses need to adopt a comprehensive strategy that involves a thorough analysis of every aspect of their supply chain [5]. This includes understanding the flow of materials from suppliers to factories, managing inventory levels, coordinating with transportation providers for timely delivery, and ensuring that products reach the end customers in the best possible condition [6]. By doing so, businesses can identify bottlenecks, inefficiencies, and areas of improvement in their logistics operations, leading to cost savings, improved customer satisfaction, and ultimately, enhanced financial performance [4].

In conclusion, logistics operations play a pivotal role in determining a business's financial performance. By investing in efficient logistics strategies and continuously refining their logistics processes, businesses can significantly enhance their profitability and gain a competitive edge in the market [2, 8]. This chapter aims to delve deeper into this topic and provide a comprehensive understanding of the relationship between logistics operations and financial performance.

It will explore various aspects of logistics operations, discuss the impact of these operations on financial performance, and provide insights into how businesses can optimize their logistics operations to improve their financial performance [5, 6]. The chapter will also highlight the importance of adopting a holistic approach to logistics management, which involves analyzing every aspect of the supply chain, from procurement and production to distribution and customer service [8]. By doing so, it hopes to provide businesses with the knowledge and tools they need to make informed decisions about their logistics operations and ultimately, enhance their financial performance [1, 4].

2. Logistics strategies

Logistics strategies are the plans and practices used by companies to manage the delivery of goods and services to their end users [6, 9]. Order processing, inventory control, warehousing, and shipping must all be coordinated in order to guarantee that the correct product is delivered at the appropriate time, location, and cost [10]. Effective logistics strategies can help businesses cut costs, improve operations, acquire a competitive edge in the market, and boost customer satisfaction [10–12].

Numerous situations and sectors can benefit from the application of logistics strategies, such as:

1. E-commerce: Since this is the most cost-effective method, logistics strategies are used in the e-commerce sector to oversee the flow of merchandise from the storage facility to the client's entrance (last mile). Transport companies use air, sea, and land transportation to ensure the timely and safe delivery of their goods.

2. Production: Logistics strategies regulate the inflow and outflow of materials and goods in the manufacturing sector. In order to keep production lines running smoothly, the aforementioned responsibilities include monitoring inventory levels, scheduling transportation, and ensuring that the right parts are available when they are needed [13–15].

3. Retail: Using logistics strategies, the basic and manageable approach makes it appropriate for the movement of products straight from the warehouse to the retail locations. It is necessary to manage transportation schedules and coordinate with suppliers in order to guarantee that products are delivered to stores on time.

4. Healthcare: Logistics plans govern how medical supplies and equipment are transported. This includes arranging for transportation, monitoring stock levels, and ensuring that medical supplies are delivered to hospitals and clinics on time.

5. Military: Logistics strategies are used to manage the movement of personnel, supplies, and equipment in the military. Organizing transportation, communicating with suppliers, and guaranteeing the availability of the necessary resources when needed are all necessary to support military operations.

2.1 Using engineering logistics to enhance financial performance

Tailored logistics engineering is a critical component in achieving financial performance targets. In order to guarantee that engineering projects are completed on time, within budget, and in accordance with required quality standards, it entails the effective management of resources, including personnel, tools, and materials [6, 15]. Careful planning and coordination are necessary for effective engineering logistics, with an emphasis on streamlining workflows, reducing waste, and increasing productivity [15, 16]. Organizations can enhance their financial performance, attain superior results, and obtain a competitive advantage in their respective markets by putting best practices in engineering logistics into practice [1, 2]. We must act proactively to lessen the impact of transportation systems about the surroundings as they develop and become more integrated with the natural world [15]. Future transportation systems must place a high priority on automation, power supply, control and management, and ecological sustainability in order to accomplish this. In order to accomplish this, the initiative is focused on three main tenets. Transport and logistics management on a large scale is suggested in **Figure 1** to increase productivity and reduce costs.

Effective engineering logistics can be applied to a wide range of engineering projects such as manufacturing procedures, transportation networks, and construction projects. In the context of a construction project, engineering logistics may entail the synchronization of material and equipment delivery to the construction site, guaranteeing their availability in the appropriate quantities and at the appropriate times [3, 17]. Managing the workforce and making sure the right people with the right skills are available when needed could also be part of it. Similar to this, engineering logistics may be used in a manufacturing process to maximize production workflows, reduce downtime, and guarantee that supplies and machinery are available when needed [2, 18]. Managing the supply chain and making sure that raw materials are sourced profitably and efficiently could also be part of it. Since the invention of computers and operational research (OR), transportation and logistics have advanced dramatically [15, 19]. But with the negative environmental effects of transportation combined with the high logistical costs for businesses, a new challenge has emerged.

A subterranean supply and freight transportation network that connects and moves between cities is known as the underground logistics system (ULS) [17, 19]. It is an eco-friendly form of transportation that runs on electricity. It can also efficiently cut expenses and traffic pressure on the roads, which eases traffic congestion in cities and lowers pollution.

As a novel approach to easing traffic congestion and cutting pollution, several nations are currently exploring the creation of urban subterranean logistics systems [18, 19]. Engineering logistics in a transportation system could include streamlining traffic, maintaining and servicing vehicles on a regular basis to reduce downtime, and optimizing routes [3, 18]. Additionally, it could entail overseeing the workforce

DOI: http://dx.doi.org/10.5772/intechopen.1004304

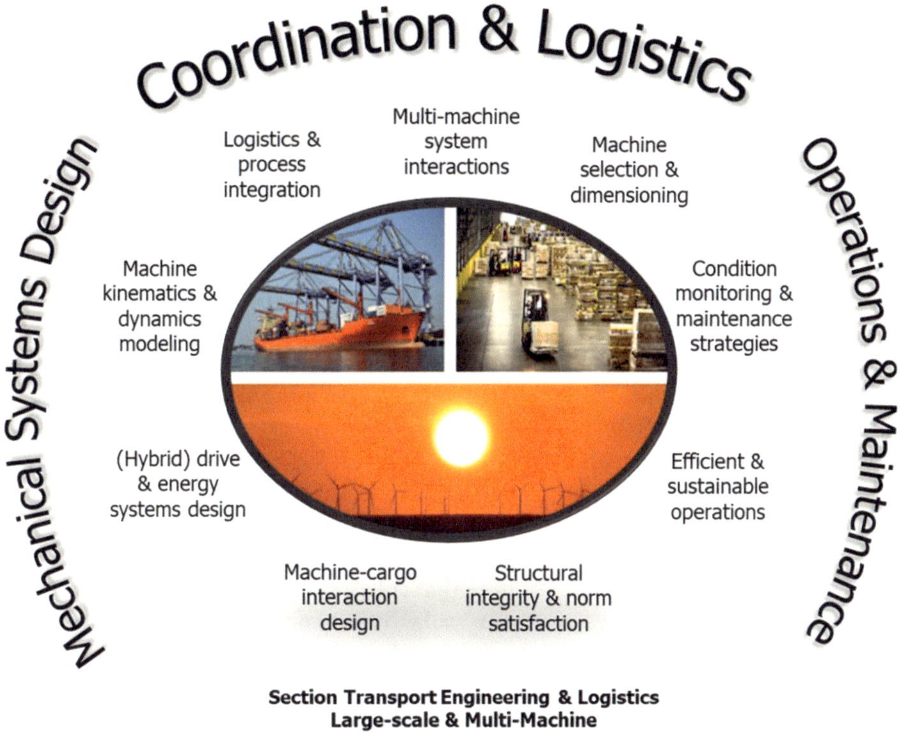

Figure 1.
Evolving transport engineering and logistics management. Source: [15].

and making sure drivers receive the instruction and resources they need to do their jobs well. – Route optimization: Assume you run a delivery business that must transport goods to several addresses throughout a city. Engineering logistics would involve optimizing the routes that drivers take in order to reduce the duration in order to counteract any potential effects and the distance required to complete deliveries [20, 21]. Identifying the most efficient routes may entail utilizing GPS and traffic information [21].

For the purpose of constructing the subterranean communications network and establishing its schedule and dynamic evolution process, a dynamic programming model has been developed [19, 21]. Using n vertices and m edges that are derived from the structural uncertainty graph of the vertices, the network can be represented as an ambiguity graph limitation [14]. The method, which consists of n logistics nodes and m tunnels, is suitable because it is not only simple but also practical. The construction of the network usually takes r years to finish. The tunnel (route) length that must be constructed each year is Dwish = Dtotal/r if the ULS's total route length is Dtotal. An analysis of the construction process over a year might look like this: Rank the logistic nodes \i, i ∈ {1, 2,...,n} according to their uncertainty measures, in decreasing order. Using the vertices structure uncertainty graph as a guide, schedule the construction of the ULS network paths in the following sequence 1, 2,..., m, where the ULS network has m edges. Create models for dynamic programming. For the dynamic programming model, the following variables are set: - k: trinnvariabilitet k ∈ {1, 2,..., m}. - dk: the distance from the kth route, or, the edge's distance BC, that is five kilometers. -ck: the shutoff.

Despite the fact that this was initially intended, the value of the kth route (ck), which connects vertices vi and vj, represents the uncertain measure φij of the current edge ek. An example of this would be the uncertain measure of edge BC, denoted by ck = 0.3. At the start of the kth stage of the ULS network, Sk + 1 represents the entire distance of all prior to. k routes (tunnels), where S1 = 0. -xk indicates whether or not the kth route has been built; xk = 0 denotes no construction; and xk = 1 indicates construction. Sk = Sk + 1 - dk * xk, where k ∈ {1, 2,..., m}, is the state transfer equation for the entire tunnel length that is permitted to be built. (8) Only the previous k routes are built using the best strategy, assuming that fk(Sk + 1) signifies the upper limit that the entire length of construction routes that are authorized can fall below Sk + 1. Given that k ∈ {1, 2, ..., m}, f0 S1 = 0, recursion in sequential order relationship is the following: fk Sk + 1 = max 0 ≤ dkxk≤Sk + 1 ckxk + fk − 1 Sk + 1 − dkxk. (9). To facilitate computation, calculo (9) is converted into the formula for the recursion relation that follows:

For every k in the interval {1, 2,..., m}, f0 S1 = 0, fk Sk + 1 = max 0 ≤ xk ≤ Sk + 1/ dek ckxk + fk − 1 Sk + 1 − dkxk. (10) The best outcome for construction routes can be achieved by sequentially solving the recursion relation Eq. (10). But a hypothetical situation is illustrated here, demonstrating how the dynamic programming model described in the text can be used. Assume that the four logistics nodes and three tunnels that make up the ULS network must be built in 4 years. Let us further assume that the length of each route (in kilometers) and the uncertain measure of each edge of the network are as follows: d1 = 10 km, c1 = 0.2 - d2 = 15 km, c2 = 0.3 - d3 = 20 km, c3 = 0.4. By applying the dynamic programming model described in the text, we can determine the best outcome for the construction routes as follows:

Logistic nodes are numbered according to their uncertainty measures' descending order of magnitude (φi): φ1 > φ2 > φ3 > φ4. ULS network paths are scheduled to become built in the following sequence: 1, 2, 3. Dynamic programming models are established: - k: trinnvariabilitet k ∈ {1, 2, 3}. - Sk + 1: the total distance of the k routes (tunnels) that were previously approved for construction in the ULS network at the start of the kth the point at which S1 = 0. -xk: indicates whether or not the kth route has been built; xk = 0 denotes no construction and xk = 1 indicates construction. Sk = Sk + 1 - dk * xk, k ∈ {1, 2, 3} is the formula for state transfer for the entire tunnel's allowable construction length. Given that fk(Sk + 1) reflects the highest possible value at which the overall distance of the construction routes that are permitted is not greater than Sk + 1, there are just the prior k routes constructed using the best possible strategy. Given k ∈ {1, 2, 3}, f0 S1 = 0, the iterated backward sequence. Relationship is listed below: fk Sk + 1 = max 0 ≤ dkxk≤Sk + 1 ckxk + fk − 1 Sk + 1 − dkxk. The recursion relation formula for Eq. (9) is changed to fk Sk + 1 = max 0 ≤ xk ≤ Sk + 1/dek ckxk + fk − 1 Sk + 1 − dkxk, where k ∈ {1, 2, 3}, and f0 S1 = 0. This makes the calculation easier.

After closely examining the data, it is discovered that the recursion relation Eq. (10) can be solved successively in the following way to find the best construction route result: - f1(S1) = 0 - f2(S2) = max{0, c1} = 0.2 - f3(S3) = max{0, c2, c1 + c2} = 0.5. As a result, the optimal result of construction routes is as follows: Year 1: Build Route 1 (length 10 km) and Year 2: Build Routes 2 and 3 (lengths 15 and 20 km). Note that the above example is merely for illustration purposes, and the actual input values and optimal results may vary depending on the particular scenario. Maintaining automobiles: Another aspect of engineering logistics is making sure that cars receive routine maintenance in order to avoid malfunctions and reduce downtime [14]. One possible approach to guarantee that your cars are always in good operating order is to plan routine maintenance inspections and repairs as needed. Aspect of engineering logistics that is not to be overlooked is workforce management [14]. In order to ensure that drivers have the abilities and resources needed to do their jobs well, this may entail hiring, training, and equipping them [3, 22]. One way to ensure safe driving practices is to train your drivers and provide them with equipment [14, 19].

2.2 Performance measures in engineering logistics organizations

To assess their operations, engineering logistics companies use a range of performance metrics. For engineering logistics companies, performance metrics are essential because they enable evaluation of the organization's progress toward its goals [17]. To evaluate its capacity to deliver goods on schedule, lower transportation costs, enhance inventory control, and boost customer satisfaction, for example, a logistics company may use performance metrics.

A metric used to assess how well a logistics company is delivering goods to customers on time is the on-time delivery (OTD) rate. This is an example of a performance measure. The quantity of goods delivered that arrived on schedule is divided by the total quantity of goods delivered to determine the OTD rate [19]. In the logistics industry, a high OTD rate means that the company is fulfilling its delivery obligations, whereas a low OTD rate implies that there are problems in the process that need to be fixed [3, 19]. A logistics company might have 100 orders to fulfill in a given month. Of them, twenty are delivered after the scheduled delivery date, and eighty are shipped out on the appointed delivery date or earlier [14]. To find the on-time delivery rate, divide the total orders that were placed. By the number of delivered orders on or before scheduled delivery date. The resultant number is then multiplied by 100 to enable a more thorough examination of certain data points using this method.

Therefore, when evaluating the responses, the numbers within each group will be taken into consideration; in this case, the OTD rate would be:

$$(85/100) \text{x} 100 = 85\% \tag{1}$$

The intriguing findings support the idea that a great way to evaluate the approach is to check if the logistics provider has an 85% on-time rate, or how well a logistics firm fulfills product delivery deadlines for customers.

The fact that the formula can be used to calculate the OTD rate is a significant strength of the study:

$$(85/100) \text{x} 100 = 85\% \tag{2}$$

Consequently, the logistics company has an 85% on-time delivery rate, which is the most pertinent conclusions (**Table 1**).

Total orders	Delivered after the scheduled time	Supplied by the scheduled date or earlier
100	85	15

Table 1.
Performance measure for on-time delivery.

Another performance indicator that is used to evaluate how quickly a logistics company is selling its stockpile is inventory turnover rate. The change in inventory rate can be found by dividing the average inventory value by the product's selling price. One example of such a business is a logistics company with a high rate of inventory turnover. By selling its inventory quickly, it was able to improve cash flow and reduce storage costs. To illustrate our point, let us consider a logistics company with annual sales of $1,000,000. Inventory turnover incidence = expense of goods sold/average inventory value can be utilized to ascertain the company's inventory turnover rate in the event that its average inventory value was $200,000 at the time. With these figures, the inventory turnover rate for the company would be $1,000,000/$200,000 = 5. It appears that this occurs five times a year for the entire inventory of the company, which are the most pertinent conclusions.

For the logistics company that sold $1,000,000 worth of goods over the course of a year with inventory turnover occurs when there is an average inventory significance of $200,000, calculated and displayed in this **Table 2**. The company's inventory turnover rate, as determined by the calculation, is five, indicating that the entire inventory is sold and replaced five times annually. It can be inferred that the logistics company achieved five times the annual sales of its entire inventory [1, 7]. This high inventory turnover rate accelerates firm inventory turnover. Effective inventory management reduces storage costs and boosts cash flow [18]. In addition, it implies that the business is making wise choices regarding production and sales, as well as efficient inventory control. Furthermore, it is suggested in Refs. [5, 18] that rethinking logistics frameworks can have a positive impact on organizational earnings by creating a comparative strategy for process integration and corporate sustainability [21].

To sum up, performance metrics are critical to engineering logistics organizations because they aid in assessing the efficiency of logistics processes. Logistics companies can pinpoint areas for improvement and decide using data to strengthen their operations by utilizing metrics, such as inventory turnover rate and on-time delivery rate [3, 5]. Organizations can accomplish their financial performance goals in a number of ways with the support of efficient engineering logistics.

First, organizations can cut costs by streamlining the flow of supplies, instruments, and machinery [7, 18]. Organizations can lower inventory carrying costs and lower the risk of stock obsolescence, for example, by implementing just-in-time inventory management techniques. Utilizing this strategy, the quantity of working capital locked up in inventory is decreased by ordering materials only as needed [4, 18, 21]. Additionally, companies can save money by preventing expensive production delays and downtime by making sure that materials are delivered on schedule and in the appropriate amounts. Along with physical products, it is suggested to include services to enhance the efficiency of the company.

Moreover, efficient engineering logistics have the potential to enhance efficiency. Organizations may focus staff on their primary responsibilities by expediting the procurement process, which will save time and effort when sourcing materials [6, 18, 23]. In addition, companies can save time on material handling and boost time on value-adding

The price of goods sold	Average value of stocks	Rate of inventory turnover
$1,000,000	$200,000	5

Table 2.
Inventory turnover rate.

tasks by making sure materials are available when needed. Faster cycle times, increased throughput, and increased efficiency can result from this [7].

Finally, high-quality engineering logistics can improve the quality of goods or services. Businesses can reduce the chance of errors and rework by making sure that materials are of the necessary caliber and are delivered on schedule [3, 6]. As a result of receiving goods or services that live up to expectations, customers may become more satisfied. Effective engineering logistics practices, for instance, may help a manufacturing company increase sales and profitability by lowering product defects and enhancing its market reputation [4].

To summarize, for an organization to achieve its financial performance objectives, efficient engineering logistics are necessary. Organizations can attain sustainable growth and boost their competitiveness by lowering expenses, increasing productivity, and improving quality [2, 3].

2.3 Performance criteria for logistics operations

Prioritizing safety while working with logistics operations requires adherence to health and safety laws, rules, directives, and other pertinent guidelines [23, 24]. The necessity of managing the flow of goods from suppliers to final consumers is a crucial component of this kind of activity [4, 21]. A supply chain risk can arise from any variation or interruption in the planned movement of materials at any point in the chain. A typical loading and offloading bay is depicted in **Figure 2**.

Unexpected circumstances can cause the flow of materials to be disrupted during their journey, which is inevitable. A progressively more serious risk to the supply chain is supply chain disruptions [21]. Operations and financial performance suffer significantly when there is intentional or unintentional disruption of the supply chain [24]. Reverse supply chain operations can be highly expensive as a result of product recalls brought on by disturbances that impair product quality [23].

Natural disasters or accidents are just two examples of the many unintentional causes of supply chain disruptions. On the other hand, deliberate disturbances might

Figure 2.
Loading and offloading bay. Source: [25].

also involve thievery, sabotage, contamination, or even a terrorist attack [24, 25]. Operations related to supply chains and logistics carry inherent risks, including those related to staffing, terrorism, labor strikes, material handling, storage, contamination, transportation delays, theft, and packaging [23]. Industrial robots are mainly used to reduce the amount of manual labor that is required [21, 23]. Robots can now complete tasks faster, safer, and more effectively than humans thanks to developments in artificial intelligence [3, 26]. Manufacturing, building, shipping, and quality assurance are some of these responsibilities. Interestingly, jobs that are too risky for humans are being completed by robots in dangerous environments. Beyond human capabilities, they are able to execute repetitive tasks with unmatched accuracy and precision (**Figure 3**) [14].

Figure 3.
Automated warehouse. Source: [25].

In order to ensure the safety and well-being of all people within an organization, including customers, guests, and employees, as well as to prevent damage to the facilities, it is imperative that effective health, safety, and security procedures are maintained. Ensuring the understanding and compliance of all relevant parties is imperative, this entails putting into place thorough and current protocols for recognizing and addressing potential health hazards, safety risks, and security threats [21]. To ensure these procedures remain effective and identify areas for improvement, continuous monitoring and assessment of these processes is also necessary [14]. Organizations can establish a secure and safe environment for all by giving priority to the upkeep of efficient health, safety, and security protocols.

It is strongly advised that you and the management team go over these procedures on a regular basis [10]. Terrorism can seriously affect the global economy and destabilize the supply chain, as demonstrated by incidents in Pakistan and events, such as 9/11. To safeguard the health and safety of everyone inside their facilities—including workers, clients, and visitors—organizations must, thus, give top priority to putting in place thorough health, safety, and security procedures. In order to ensure compliance, this entails recognizing potential risks and hazards, taking appropriate action, and teaching all parties involved in the process [4, 10]. In order to preserve the efficacy of these protocols, ongoing observation and evaluation are imperative in order to pinpoint opportunities for enhancement [21]. Organizations can establish a secure environment for everyone by placing a high priority on security, safety, and health (**Figure 4**).

In addition to having a negative effect on the economy, law and order issues can also seriously harm the supply chain sector by causing disruptions, harm, and attacks on the movement of goods, which can impact the chain as a whole [24]. The supply

Figure 4.
Supply chain network disruption 9/11.

chain and logistics processes are likewise susceptible to major effects from natural disasters. The recent conflicts, such as the Israel-Gaza war and the Russia-Ukraine war, have significantly damaged and disrupted the supply chain network, as have the earthquakes in Turkey and the floods in Libya [24]. To mitigate the impact of these incidents, appropriate risk management plans must be implemented [9, 24]. However, because the study was retrospective in nature, follow-up was not feasible.

Law and order issues can seriously harm the supply chain sector by causing disruptions, harm, and attacks on the movement of goods, which can impact the chain as a whole [24]. Despite this, a number of issues with the results persist. Mainly, this is because these issues can also negatively impact the economy [9]. The supply chain and logistics processes are likewise susceptible to major effects from natural disasters [24]. The recent conflicts, such as the Israel-Gaza war and the Russia-Ukraine war, have significantly damaged and disrupted the supply chain network, as have the earthquakes in Turkey and the floods in Libya [21]. In order to lessen the effects of these occurrences, it is crucial to have suitable risk management strategies in place [3, 24]. If there are any issues within your control, make sure to resolve them quickly and decisively. If you are having trouble solving any problems on your own, get advice and assistance. Always prioritize safety and ask for assistance rather than endangering others or yourself.

2.4 Key rules and protocols for logistics engineering

The comprehension of standard regulations, guidelines, and procedures in their entirety is essential for a logistics engineer. The following knowledge domains must be possessed in order to do this

1. Getting access to health and safety regulation information: In order for businesses to remain competitive in the market today, they must be aware of the most

recent technological advancements that can be used in logistics operations. Automation, AI, blockchain, and the Internet of Things are just a few examples of the technological advancements that have significantly impacted the logistics sector [7]. Based on the information provided, it is possible for businesses to ensure a safer, more cost-efficient, and more efficient supply chain network by using these technologies to streamline their logistical processes [3]. Transportation, order processing, inventory management, and other repetitive and time-consuming tasks can all be automated with the use of automation technology [9, 27]. Businesses produce more, require less manual labor, and make fewer mistakes as a result. Conversely, real-time tracking [24], logistical operation optimization, and predictive analytics can be achieved through the application of artificial intelligence. Organizations can anticipate demand, improve delivery times, and identify possible risks by utilizing this technology [6, 17].

The most popular strategies used today to achieve this include those that leverage blockchain technology, which allows for safe and transparent product tracking all the way through the supply chain [23]. Because blockchain reduces the likelihood of fraud, theft, and counterfeiting, it has become an especially appealing option for an organization looking to improve the visibility, traceability, and accountability of goods [4, 6]. Furthermore, by linking numerous devices and sensors, companies can use the World Wide Web (WWW) to provide real-time data and insights that enhance logistics operations. Accepting new technologies calls for a methodical and cautious approach, which needs to be recognized. This technology has become apparent in recent decades [3, 17]. This has only recently become a very interesting phenomenon for organizations, which need to evaluate their current infrastructure, weigh the advantages and disadvantages, and make sure employees have the necessary training before introducing new technology [6]. Because companies who optimize their logistics operations can boost the efficiency and security of their supply chain network, this has become a very popular area recently.

2. Methods and protocols for logistics engineering activities: One of the most significant conclusions is that, in order to guarantee the effective execution of logistics operations, a thorough comprehension of the pertinent planning and scheduling methods and protocols is required [4, 6]. Being fully aware of the procedures followed when moving people, supplies, and products is part of this. To handle the different phases of logistics operations, from the first planning phase to the last delivery, a thorough understanding is required [3, 27, 28]. The solution this work offers is its main contribution: When logistics operations are performed as effectively as possible, there is a decreased chance of mistakes or delays because pertinent protocols and procedures are understood [4]. Detailed knowledge of the protocols and guidelines involved is therefore required for the effective planning and scheduling of logistics, and these are the contributions of this work.

3. The advantages and restrictions of transportation. It follows that in order to ensure efficient logistics operations, a full grasp of the advantages and constraints of each mode of transportation is required [9, 10]. A few issues remain, so let us examine each mode of transportation in more detail. It is crucial to discuss fundamental concepts first.

DOI: http://dx.doi.org/10.5772/intechopen.1004304

Air transportation is a great option for goods that need to reach their destination quickly because of its reputation for dependability and speed, but gathering enough data can be a challenging process [23]. Its highest cost can be a drawback for businesses transporting large quantities of cargo though.

Road transportation: The most common type of transportation is dependable, flexible, and easily accessible. These are built for a variety of reasons. For smaller shipments, it is cost-effective and ideal for short- to medium-distance transportation [9]. On the other hand, bad weather and traffic congestion can cause delays. The most cost-effective mode of transportation for bulk commodities being transported over large distances is rail [7, 23]. It is also environmentally friendly and has the least carbon footprint. However, it can only travel on specific routes and to specific locations, and its overall speed is slow [29]. Long-distance bulk goods shipping is an excellent fit for maritime travel because it is the most economical and cargo-capable mode of transportation [6, 10, 17]. The primary achievements, including contributions, can be summed up as follows despite being somewhat slow and having erratic shipment schedules [29]. Making educated decisions that can optimize logistics operations requires an understanding of the benefits and drawbacks of each mode of transportation.

4. Choosing the mode of transportation: To make sure that your choice of transportation is well-informed, there are a number of factors to take into account as several other criticisms have been made [6, 10]. The journey's objectives, time constraints, distance to be traveled, cost of transportation, and consideration of the most suitable mode of transportation are two major criticisms frequently leveled at this approach [29]. This demonstrates the potential utility; you can weigh the advantages and disadvantages of each form of transportation and select a reasonable, cost-efficient, and safe option by considering these factors.

5. Transport provider evaluation criteria: This approach is available in multiple versions because there are a lot of factors to consider when selecting a transportation provider [7, 9]. These include the type of cargo being transported, the distance to be covered, the anticipated delivery time, the cost of transportation, and the reliability of the offering, as well as the level of security offered. These techniques can also be combined with the ones that evaluate the provider's overall reputation for providing excellent customer service [4, 6, 10]. It is also important to consider how well-equipped they are to handle unanticipated events, such as inclement weather or delays. By carefully [10] weighing these variables, you can select a carrier that best meets your needs and ensures the timely and safe delivery of your goods [29]. These methods are represented by a variety of different approaches.

6. Order scheduling: A comprehensive method was created and applied to ascertain how to schedule orders in an efficient manner that complies with organizational procedures. To do this, it is necessary to fully comprehend all relevant processes. Identifying the optimal time to fulfill, obtaining the data needed for each order [7], and assessing the resources needed are typical steps in this process that are evaluated using this common technique [4, 9]. This has historically been the preferred approach, but there has been a recent shift in the awareness that there are a number of other factors to take into account when scheduling orders, including

the availability of equipment [10], the level of expertise of the workforce, and the lead time required for raw material delivery [29]. Order scheduling has historically been made simpler with the use of this information, resulting in higher output and happier customers [9].

7. Provision of information and documentation: Understanding the procedures that an organization follows in relation to providing information or documentation is essential. Understanding who needs to receive information and in what format is necessary for this [27]. Essential skills include knowing where to find relevant documents or information and ensuring that the appropriate people see it [29]. This requires a thorough understanding of the various roles and responsibilities within the organ in addition to the needs and expectations of the individuals who will be receiving the information or documentation [9]. Organizations can make sure they are fulfilling their responsibilities and offering a high standard of service by having the ability to deliver information or documentation that is clear, concise, and customized to the needs of various audiences.

8. Correct documentation issues: In many professions, including law, medicine, and business, it is crucial to issue documentation accurately. Securing the proper documents at the appropriate time, in the appropriate order [9], and to the appropriate individuals is a more involved process than just issuing them. In order to guarantee that the information provided is clear, documentation must be issued correctly [27]. Mistakes, misunderstandings, and misinterpretations can be prevented with proper documentation. In data and information, it also aids in maintaining accuracy and consistency [4, 30]. Issued documentation correctly guarantees compliance with legal and regulatory requirements in addition to maintaining accuracy and consistency [17]. In the event that noncompliance occurs, it helps to prevent legal conflicts, fines, and other legal consequences. In summary, companies, organizations, and individuals must all function effectively and efficiently while minimizing the risks associated with incorrect documentation [10]. This requires an understanding of the significance of correctly issued documentation.

9. Document storage: Not enough emphasis can be placed on how important it is to properly store documents in any organization. Carefully following organizational policies and paying attention to detail are necessary for the completion and security of documentation [16]. In addition to being familiar with the rules and processes for organizing and maintaining documents, this also entails understanding the various protocols for document retrieval when necessary [9]. To protect sensitive information from unauthorized access, it is also essential to make sure that documents are stored securely.

Organizational efficiency can be greatly increased by a well-performing document storage system that facilitates easy access to information and streamlines processes. Reducing the possibility of misplacing crucial documents that could have significant effects on the company is another benefit [17]. Another crucial component of document storage is regulatory compliance, which can be guaranteed by an efficient system that makes sure the company complies with all applicable laws. In general, an essential part of any organization's operations is appropriate document storage [4, 27]. Ensuring that documents are properly organized, maintained, and stored can yield substantial advantages for the company, but it does require a careful approach.

10. Issues with goods transportation: It is crucial to plan ahead and have a thorough awareness of all possible potential problems that could occur when moving goods [7, 27]. Numerous factors, ranging from unanticipated events to logistical challenges, can affect the successful delivery of goods. For example, unforeseen traffic or weather conditions can cause delays in transit, which can result in late deliveries and disgruntled customers [7, 16]. Comparably, inadequate packaging or careless handling by the shipping company may cause damage to goods during shipment, incurring additional expenses for replacing or repairing the damaged items.

Additionally, inadequate or missing documentation can result in serious hold-ups at customs and significant delays. For companies that depend on prompt item delivery to satisfy client demand and revenue goals, this can be especially troublesome [7]. Incorrect labeling or declarations are examples of customs-related problems that can occur and result in extra charges and penalties [29]. The first step in addressing these possible issues [4, 27] and reducing their negative effects on your company is being aware of them. Knowing how to deal with these issues is also crucial. Establishing backup plans, for instance, can facilitate prompt response and customer communication in the event of unforeseen delays or damage [7, 29]. Analogously, putting in place a system to guarantee appropriate labeling and documentation can help you completely avoid problems connected to customs.

11. Data requirements for logistics operations scheduling: Organizing logistics is a complicated process that necessitates a thorough comprehension of the various kinds and sources of data needed to satisfy client demands [7, 30, 31]. Information regarding stock levels, routes taken by vehicles, delivery schedules, and resources that are available are just a few examples of what is included here [16]. It is essential to have precise and current information on hand in order to plan logistics activities efficiently. Many methods, including speaking with suppliers, clients, and transportation companies [8], as well as utilizing technology and data analysis tools, can be used to obtain this information. Businesses can guarantee that they are satisfying customer demands in a timely and effective way, using their own resources as efficiently as possible [27], and keeping costs down by having a thorough grasp of the information requirements for scheduling logistics operations.

Worked example: Let us say a company has to ship goods to different states in the nation. The company must be aware of the various information sources and types needed in order to plan logistics activities efficiently. Information regarding product inventory levels, available transportation routes, delivery schedules [7], and transportation resource requirements are all included in this. Communicating with suppliers, clients, and transportation providers can help the company obtain this data [9, 31]. In order to evaluate the information and come to wise decisions, they can also make use of technology and data analysis tools. A business can guarantee that it is satisfying customer demands in a timely and effective way, optimizing its own resources and cutting costs [32] by having a thorough understanding of the information requirements for scheduling logistics operations.

To provide you a better idea of the significance of information requirements for logistics operation scheduling, let us go into more detail about the worked example below. In general, the issue can be resolved by Suppose that the business in the example has a substantial amount of inventory that needs to be shipped across the country [7]. To ensure that logistics operations meet customer demands, the simplest method is to schedule them effectively [16, 17]. Their inventory levels may be a factor in solving this challenging issue; however, in order to do so, they must be aware of the demand for their products in each location. Numerous strategies to deal with this problem have been put forth; with this data, they can decide more intelligently which products to ship and in what quantities [4].

A more comprehensive approach has recently been put forth, entailing the requirement for details regarding the cost, duration, and length of every one of the options for transportation [22]. Utilizing this data to identify the most successful and economical method of product delivery has been one approach taken to solving these issues [8, 32].

The aim of the endeavor is to eradicate the existing obstacles in obtaining data about any unique circumstances or restrictions that might be relevant to specific regions along with the delivery timelines that their customers require [8]. Utilizing this data [16], they can make improvements by guaranteeing that their products are delivered punctually and in accordance with any relevant laws or regulations. The current general consensus is that the resources needed for transportation, such as cars, drivers, fuel, and upkeep, should be the least of their concerns [6, 32]. Having more knowledge, they can use this information to ensure they have the resources needed to transport their goods efficiently [29]. Given the potential gravity of the situation, it is critical to gain a thorough understanding of all the information needed [31]. With this understanding, the business can efficiently plan logistics operations, promptly and effectively meet customer demands, optimize its own resources, and cut expenses [22, 32].

12. Logistics operations organizational systems and procedures: It is common knowledge that comprehension of the systems and processes used to arrange logistics activities is essential to meeting customer expectations [29]. There is currently a general consensus that the systems and procedures used in logistics scheduling can be influenced by the industry, the type of transportation, and the needs of the customer. A number of well-liked methods and structures consist of:

    a. To get around some of the innate restrictions, the transportation management system (TMS) was introduced. TMS, a software platform, is used by logistics companies to plan, execute, and optimize the movement of goods [30]. This is presently being considered.

    b. Systems for enterprise resource planning (ERP): ERPs combine different business operations, such as logistics, to provide a centralized system for operations management.

    c. Lean logistics is an approach that maximizes the effectiveness of logistics processes by getting rid of waste and increasing productivity.

    d. Six Sigma: To increase quality and efficiency in logistics operations, Six Sigma is a data-driven methodology.

Day	Time	Activity
Monday	9:00 AM	Pick up goods from warehouse
Monday	12:00 PM	Deliver goods to retail store A
Tuesday	9:00 AM	Pick up goods from warehouse
Tuesday	12:00 PM	Deliver goods to retail store B
Wednesday	9:00 AM	Pick up goods from warehouse
Wednesday	12:00 PM	Deliver goods to retail store C
Thursday	9:00 AM	Pick up goods from warehouse
Thursday	12:00 PM	Deliver goods to retail store D
Friday	9:00 AM	Pick up goods from warehouse
Friday	12:00 PM	Deliver goods to retail store E

Table 3.
Weekly delivery schedule.

Take into consideration a logistics company that has to plan the movement of goods from a warehouse to a retail location in order to provide an example of scheduling in a table format [29]. A weekly schedule example is provided in **Table 3**.

Finding out what proportion of the vehicles' time is spent being used during operating hours is necessary in order to compute the vehicle utility rate. Assuming 2 hours for each retail store, we can calculate the time needed for both product pickup and delivery.

It follows that each car can make a trip to a retail location every 4 hours, which includes 2 hours for travel time and 2 hours for pickup and delivery. Every vehicle has access to 6 hours of operating time if the operating hours are from 9:00 to 15:00. In light of this, each vehicle has a 4-hour cycle that it can finish in three different times: at 09:00, 13:00, and 10:00 or 14:00. This equals three daily trips for each vehicle or a total of six trips for both.

Thus, the following formula would be used to determine the vehicle utility rate.

$$\text{Vehicle utility rate} = \left(\text{Total time vehicles } \textit{are} \text{ utilized} / \text{Total operating time}\right) \text{x } 100 \quad (3)$$

$$\text{Total time vehicles } \textit{are} \text{ utilized} = 6 \text{ trips x } 2 \text{ hours } \textit{per} \text{ trip} = 12 \text{ hours} \quad (4)$$

$$\text{Total operating time} = 2 \text{ vehicles x } 6 \text{ hours } \textit{per} \text{ vehicle} = 12 \text{ hours} \quad (5)$$

$$\text{Vehicle utility rate} = (12/12) \text{ x } 100 = 100\% \quad (6)$$

In light of this, the two vehicles' vehicle utility rates are 100%.

13. Issues with logistics scheduling: In the field of logistics, logistics scheduling is a complicated and multidimensional procedure [7]. Planning the movement of goods, controlling inventory levels, and liaising with suppliers and customers are just a few of the many tasks involved. It is not always easy going through this process, though. When logistics operations are scheduled to meet customer

requirements, a number of potential issues could arise [8, 31]. These problems can include equipment malfunctions, supply chain interruptions, and unforeseen delays and bottlenecks.

Logistics managers need to be extremely knowledgeable about potential problems that may come up when planning logistics operations in order to effectively handle these difficulties [22]. They must possess the ability to recognize possible issues before they arise and create workable plans of action to lessen their effects. This could entail creating backup plans and systems in case of emergencies, establishing channels of communication to notify all parties involved [22], and putting technology to use in order to increase productivity and streamline processes. Logistics managers can guarantee timely and minimally disruptive fulfillment of customer requirements by adopting a proactive scheduling strategy for their operations [6, 31]. Long-term competitive advantage and business growth are also fueled by this in addition to helping to foster customer trust and loyalty.

Several issues may arise during the scheduling of logistics operations, such as:

a. Delays: There are a number of reasons why delays can happen, including traffic, bad weather, malfunctioning equipment, or unplanned circumstances. These delays have the potential to completely halt logistics operations, which would have an impact on the entire supply chain.

b. Bottlenecks: Capacity or resource shortages that prevent the flow of goods or services at the desired volume can result in bottlenecks. This may result in traffic jams and delays, which would raise expenses and decrease productivity.

c. Supply chain disruptions: There are several causes of supply chain disruptions such as unforeseen shifts in supply or demand, natural disasters, and political upheaval. These hiccups may have a major effect on logistics operations, leading to extra expenses, delays, and lost cargo.

d. Communication breakdowns: When various parties involved in logistics operations do not coordinate and collaborate with one another, communication breakdowns may occur. This may result in miscommunication, a lag in decision-making, and a lack of transparency regarding the state of logistics operations.

e. Erroneous demand or supply forecasting can result in either an excess or a shortage of inventory, raising expenses and decreasing the effectiveness of logistics processes.

14. Timings and deadlines for logistics operations setting up expectations with clients for precise timings and deadlines is essential to delivering excellent logistics services: The details of the operation [22], such as the pickup or delivery date and time, the transit schedule, and any other pertinent information, must be discussed. It is possible to guarantee that your clients are well-informed and happy with the services they receive by settling on these aspects in advance. Knowing how to handle these conversations professionally and skillfully while accounting for any unforeseen difficulties or adjustments that might occur during the procedure is also crucial.

DOI: http://dx.doi.org/10.5772/intechopen.1004304

15. Surveillance of the logistics schedule: The logistics schedule needs to be closely monitored in any operation where resources or goods are being transferred [33]. Due to the potential for serious consequences such as missed deadlines, increased costs, and even lost revenue, the logistics schedule is essential [31]. To avoid such circumstances, it is imperative to regularly monitor the logistics schedule and create backup plans as needed. This requires an understanding of the intricacies of the logistics schedule in addition to the ability to identify potential bottlenecks or obstacles and adjust plans appropriately [33]. Through this approach, an organization can guarantee the efficient and profitable functioning of its logistics processes, thereby augmenting its triumph and earnings.

Today's fast-paced world depends on logistics for any business operation involving the movement of goods or resources [30]. With time, the concept has evolved. A typical explanation is that the supply chain's foundation is its logistics, and any disruptions or delays in the logistics plan can have a big effect on the whole business [31, 32], including the production and retail location [13]. For ease of illustration, suppose a manufacturer receives an order to manufacture and deliver a specific product to a customer within a given time frame. There is a possibility that delays could happen anywhere in the supply chain, including during the shipment of finished goods or raw materials, if the logistics schedule for this order is not properly tracked [13, 33]. This would boost performance. As a result of these delays, the production process may be negatively impacted, missing deadlines and perhaps incurring penalties for late deliveries [28].

So, presuming this, the following techniques can be used to correspondingly ensure that a retailer using a just-in-time inventory system keeps a close eye on its logistics schedule in order to avoid stockouts or overstocking [31]. For this reason, stockouts are frequently equated with a retailer's incapacity to meet customer demand in the event that there is a delay in the delivery of goods [13]. In contrast, overstocking could lead to unnecessary expenses and costs associated with keeping inventory. It is crucial to regularly check the logistics schedule and create backup plans as needed to prevent such situations. To do this, one must be able to comprehend the complexities of the logistics schedule, spot possible bottlenecks or roadblocks, and modify plans as necessary [32]. This can involve having backup suppliers or manufacturers, alternate routes or forms of transportation, or even extra inventory on hand. In conclusion, in Ref. [33] the success of any business operation involving the transfer of resources or goods depends on keeping an eye on the logistics schedule and developing backup plans. Through this approach, entities can guarantee the seamless and effective functioning of their logistics processes, resulting in increased prosperity and financial gain [13].

16. Constant evaluation and analysis of logistics operations: It is critical to periodically examine and assess your logistics operation to pinpoint areas for improvement in the fast-paced business climate, which exists today. You may fulfill your clients' shifting needs and remain one step ahead of the competition by doing this. Increasing efficiency, effectiveness [33], and customer satisfaction can be attained by consistently refining your logistics operations.

Your logistics operations need to be improved in a number of ways. Simplifying your transportation routes is crucial if you want [33] to cut down on delivery costs and times. Accurate demand forecasting, fewer stock-outs, and less overstocking are further ways to improve inventory management [31]. Furthermore, improving departmental coordination and cooperation can help you cut down on needless delays and achieve better coordination [13]. Another crucial area to concentrate on is cutting waste and inefficiencies. Reducing idle time, minimizing errors, and identifying and removing bottlenecks in your logistics processes are some ways to achieve this. Utilizing technology is also crucial for increasing tracking and visibility, which enables you to oversee and control your logistics processes in real time. You must undertake a thorough analysis of your present logistics procedures in order to pinpoint areas that can be improved [12]. Once the areas that require improvement have been determined, you can create a plan to address them and endeavor to achieve more effective and efficient logistics operations.

17. Logistics equipment: It is essential to comprehend the equipment required for logistics. It is critical [33] to understand, which equipment is suitable for receiving, storing, and shipping goods, as well as how to choose the right equipment. This entails knowing what equipment is needed at each stage of the process and selecting the best equipment for the task [12]. A novel approach is put forth to tackle this phenomenon that has remained unexplained thus far, positing that material handling in logistics has undergone significant transformation ever since the Fourth Industrial Revolution began to emerge [13]. Advanced technologies, such as automation and robotics, in addition to the Internet of Things (IoT), have improved effectiveness, dependability, as well as efficiency of logistics operations [31]. Several examples of how the Fourth Industrial Revolution has changed logistics and material handling are provided here:

    a. Automated guided vehicles (AGVs): These are autonomous vehicles that can move goods and materials throughout a production line or warehouse [18]. They can navigate and avoid obstacles thanks to the sensors and software that are installed in them.

    b. Drones: Products are delivered to isolated or difficult-to-reach areas using these unmanned aerial vehicles or UAVs. They are especially helpful in the logistics sector for prompt and effective delivery of small packages.

    c. Robotics: In logistics, robots are frequently utilized for material handling operations such as order picking, depalletizing, and palletizing. They are a great option for repetitive tasks because they are highly accurate and can work nonstop.

    d. Augmented reality (AR): AR technology helps logistics companies run more accurately and efficiently in their warehouses. Workers can locate goods and materials more quickly by using AR glasses, which allow them to see digital images superimposed on the real world.

These are but a few illustrations of the ways that the Fourth Industrial Revolution has changed logistics and material handling [18]. We can anticipate even more creative solutions to appear in the logistics sector as long as technology keeps developing.

18. Growing concern: The rising transportation of goods in logistics operations has raised significant concerns regarding the environment. With this increase, as a

result, the environmental impact has increased accordingly [11, 12]. Common environmental issues within logistics operations encompass air pollution, water pollution, noise pollution, and greenhouse gas emissions.

Hazardous gases, such as carbon monoxide and nitrogen oxides that are released from moving automobiles, are what cause air pollution. Waste products from moving automobiles can leak into water sources and contaminate them [12]. Vehicles that produce loud noise are the source of noise pollution, which can have a detrimental effect on the health and happiness of those who live and work nearby [11]. A significant factor in climate change is the emissions of greenhouse gases, mainly carbon dioxide. Transportation vehicles energy consumption in warehouses and other facilities are two major sources of greenhouse gas emissions that are produced by logistics operations [18].

Logistics companies can employ diverse tactics to tackle these environmental concerns [12]. Using fuel-efficient cars or moving to alternative fuels, such as biofuels, hydrogen, or electric power, are two ways. Enhancing logistics operations' effectiveness is an additional tactic that can lower emissions and energy usage. Furthermore, businesses can implement sustainable measures, such as waste reduction and recycling. Logistics processes can become more sustainable and friendly to the environment by doing this.

19. Techniques for minimizing environmental Impact: It is critical to comprehend the techniques that logistics operations can employ to lessen their negative environmental effects. Various techniques can be employed to mitigate the influence on the environment, such as:

    - The simplest and most efficient way to lessen your influence on the environment is to reduce, reuse, and recycle. It entails cutting down on the amount of raw materials used, reusing items, and recycling materials to reduce waste.
    - Usage of renewable energy sources: Using renewable energy sources can help lower the carbon footprint of logistics operations. Examples of these include solar, wind, and hydropower.
    - Reduce fuel use and emissions: Transportation optimization can lead to reduced emissions and fuel consumption. This can be accomplished by streamlining delivery routes and utilizing more fuel-efficient modes of transportation, such as hybrid or electric cars.
    - Adopt sustainable packaging: Using biodegradable or compostable packaging materials can help minimize waste production and the environmental impact of logistics operations.
    - Adopt green storage practices: The installation of motion sensors and energy-efficient heating and lighting systems are examples of green warehousing practices that can help minimize the environmental impact of logistics operations and reduce energy consumption.
    - Through the implementation of these techniques, logistics companies can considerably mitigate their ecological footprint and foster a more sustainable future.

20. Using current and approved policies and procedures: Organizations must make using current and approved policies and procedures a top priority in order to protect the safety and well-being of everyone who is present in their facilities, including workers, clients, and visitors. Organizations can recognize possible risks and hazards, take appropriate action, and stop any bad effects by following these policies and procedures.

Knowing the consequences of disregarding these guidelines and procedures is equally crucial though. A number of unfavorable effects, from small interruptions to the supply chain network to significant repercussions such as accidents, injuries, or even fatalities, can spring from such noncompliance [20]. The significance of following current policies and procedures, as well as the possible consequences of not doing so, must thus be emphasized to employees by organizations [34]. To guarantee the efficacy of the established policies and procedures and to pinpoint areas in need of improvement, ongoing evaluation and observation are essential. Organizations may establish a safe environment for everyone and reduce risks and hazards by giving priority to the application of current policies and procedures [12].

21. Innovations in technology for logistics operations: To remain competitive in today's ever-evolving market, firms must be aware of the most recent developments in technology that can be used for logistics operations. Automation, artificial intelligence, blockchain, and the Internet of Things are just a few examples of the technological advancements that have significantly changed the logistics industry [18]. In actuality, it can be handy for businesses to use these technologies to streamline their logistics procedures and guarantee a more effective, affordable, and safe supply chain network.

Automating time-consuming and repetitive tasks, such as order processing, transportation, and inventory management, can be achieved, at least in part, through automation technology [20]. As a result, businesses experience decreased errors, lower labor costs, and higher output. In contrast, artificial intelligence can be applied to real-time tracking, predictive analytics, and logistics operation optimization [11, 18]. The first step of the method assumes that companies can use this technology to predict demand more precisely, identify potential risks, and expedite delivery. Blockchain technology offers a transparent and secure way to track goods throughout the supply chain network. Blockchain improves the visibility, traceability, and accountability of goods for organizations [35], thereby lowering the risk of fraud, theft, and counterfeiting. Moreover, enterprises can harness the Internet of Things (IoT) to improve logistics operations by connecting various devices and sensors, thereby supplying real-time data and insights [18].

However, it is imperative to keep in mind that implementing new technologies must be done so cautiously and strategically. It is imperative for organizations to assess their existing infrastructure, weigh the benefits and drawbacks of incorporating new technologies, and ensure that personnel have been properly trained to utilize them [11]. Creating a more dependable and secure supply chain network and optimizing their logistics procedures are two ways that organizations can achieve this.

22. Responsibility and reporting: It is critical to comprehend your reporting and accountability duties for any organization to function efficiently. Understanding whom to report issues to and being aware of your full responsibilities are

essential. Recognizing potential issues and promptly reporting them to the proper higher authorities are necessary for accomplishing this [35]. Ensuring that the appropriate individuals are aware of these issues and can respond accordingly also requires proper communication of these issues at the appropriate time and place. If you know exactly what your reporting and accountability responsibilities are, you can ensure that your business runs smoothly and successfully.

3. Conclusion

This text concludes a chapter that examined the relationship between financial performance and logistics operations. Below, you will find important details and supporting information:

Due to their influence over the profitability and efficiency of the different tasks involved in transporting goods and services from point of origin to point of destination, logistics operations are, therefore, critical in determining the financial performance of a company [11, 34]. The purpose of this is to offer solid proof that businesses can greatly enhance their financial performance by streamlining these tasks.

Optimizing logistics operations yields several advantages:

- lower transportation and storage expenses.
- increased revenue as a result of prompt customer delivery of the goods.
- increased contentment among clients.
- enhanced inventory management.
- improved storage space utilization.

Still, there are a lot of obstacles to overcome and room for advancement in logistics practices [11]. For instance, since it enables businesses to foresee client demands and make appropriate plans, demand forecasting is a crucial competency for efficient supply chain management [36]. Whether or not this was the main cause of manufacturers' ability to minimize inventory costs, cut waste and obsolescence, optimize transportation and logistics operations, and raise overall supply chain efficiency is difficult to determine [12, 18]. This is the best course of action that lowers expenses and boosts revenue.

It is obvious that these experiments have significant positive effects. The creation of a logistics business continuity plan, which can assist organizations in handling unforeseen disruptions and disasters that may have an influence on their logistics operations, is another area that needs to be given optimal parameters [11, 36]. These technologies have a lot of promise. By strengthening their disaster immunity and logistics competitiveness, businesses can increase their operational and financial performance by creating a logistics business continuity plan [34].

To summarize, the enhancement of a company's financial performance is contingent upon the implementation of logistics operations, which necessitate ongoing education and adjustment to the ever-changing business landscape [20, 35].

This chapter presents a more intriguing and realistic scenario in which businesses can invest in effective logistics strategies and continuously improve their logistics processes to not only increase their profitability but also gain a competitive advantage in the market.

The efficiency and profitability of the numerous activities involved in transporting goods and services from point of origin to point of destination are impacted by logistics operations [34], which are ultimately critical to a company's financial performance [12]. The key finding of this research is that businesses can greatly enhance their financial performance by streamlining these tasks.

- One or more of the following are advantages of effective logistics operations;
- lower costs for storage and transportation;
- more income from on-time product delivery to clients;
- better inventory control;
- and more economical use of warehouse space.

Still, there are a lot of obstacles to overcome and room for advancement in logistics practices. For instance, since it enables businesses to foresee client demands and make appropriate plans, demand forecasting is a crucial competency for efficient supply chain management [11]. In general, the findings suggest that manufacturers can minimize inventory holding costs, minimize waste and obsolescence, optimize transportation and logistics activities [18, 37], and enhance the overall efficiency of the supply chain by accurately predicting demand. The inference that this results in lower costs and higher profitability is supported by these data [12].

Another area that requires attention is the development of a logistics business continuity plan, which can help businesses to cope with unexpected disruptions and disasters that may affect their logistics operations [34]. By developing a logistics business continuity plan, businesses can improve their operational performance and financial performance by increasing their logistical competitive capability and enhancing their disaster immunity.

In conclusion, logistics operations play a vital role in enhancing a business's financial performance; however, they also necessitate ongoing education and adjustment to the ever-changing business landscape [11, 35]. This data lends credence to the idea that businesses can boost profitability and gain a competitive edge in the marketplace by making strategic investments in efficient logistics plans and ongoing process improvement [25].

Insufficient research has been done on the relationship between financial performance and logistics operations. There are still many areas that need to be investigated further, even though numerous studies have looked at how different aspects of logistics operations affect financial performance [34, 37]. For instance, there is still a lack of knowledge regarding the capabilities of digital technologies, and how they affect logistics operations across various industries and contexts. International trends, such as e-commerce and sustainability, also have an impact on logistics operations, but their effects on financial performance are not sufficiently considered [12, 25].

These issues are important for developing effective and efficient logistics strategies that can increase profitability and competitiveness in the dynamic business environment.

Investing in successful logistics strategies and continuously enhancing their logistics procedures can help businesses boost profitability and acquire a competitive edge in the market [11]. Strategies for logistics that are effective are those that maximize the effectiveness and efficiency of different tasks associated with moving goods and services from the point of origin to the destination.

Monitoring, assessing, and improving the performance of logistics activities such as transportation, inventory control, storage, and packaging is what is meant by continuous improvement of logistics processes [12, 28].

Through the implementation of efficient logistics strategies and the ongoing enhancement of logistics procedures, businesses can boost their earnings and secure a competitive edge within their industry:

- Minimizing waste, optimizing routes, and leveraging digital technologies are key strategies for reducing inventory and transportation costs.

- Ensuring timely delivery of goods to customers and increasing customer satisfaction will in turn boost sales and create value-added services.

- Streamlining workflows, automating tasks, and coordinating with suppliers and partners will improve operational efficiency.

- By developing emergency plans, mitigating incidents, and complying with regulations, risk management can be enhanced.

- Additionally, promoting sustainability can be achieved through reducing carbon emissions, conserving natural resources, and adopting environmentally friendly practices.

This chapter aims to give readers a thorough understanding of the connection between logistics operations and financial performance. Additionally, businesses aiming to enhance their logistics capabilities are supposed to find it a helpful resource. Transportation of goods and services from a point of origin to a destination is referred to as logistics operations [18]. This covers duties such as packing, warehousing, transportation, and inventory control [15, 25]. The network of businesses and operations that creates and delivers goods and services to consumers is known as the supply chain, and these jobs have an effect on its efficacy and efficiency [28]. How well a business generates and manages its financial resources is determined by its financial performance. It comprises metrics, including market share, cash flow, profits, sales, and return on assets.

The ability of a business to generate value for its constituents—including clients, investors, and staff—is reflected in these metrics. The premise that more successful and efficient logistics operations can result in higher financial performance underlies the relationship between financial performance and logistics operations [15, 28]. Through the optimization of logistics processes, businesses can lower expenses [34], boost revenue, and improve customer satisfaction [18, 35]. A company's financial performance may increase as a result of these outcomes, and it may gain a market edge. It is not evident or easy to understand how financial performance and logistics operations are related [11, 25]. The kind of industry, the size of the business, the kind of goods and services offered, the traits of the clientele, and the state of the market are just a few of the variables that influence it [37]. The subtleties and complexity of the relationship between financial performance and logistics operations must, therefore, be understood.

This chapter's goal is to give readers a thorough understanding of how logistics operations and financial performance are related. Examined are the terms and definitions of financial performance and logistics operations [15], the variables and factors that affect the relationship, the models and measurement techniques that assess the relationship, and the tactics and approaches that maximize the relationship. Furthermore, case studies and examples of how businesses have successfully enhanced their financial performance through efficient logistics operations are provided.

Additionally, this chapter is meant to be a helpful resource for businesses trying to enhance their logistical capabilities. It offers analysis and suggestions for creating, putting into practice, and assessing logistics plans that can raise bottom line results. Notably from Ref. [25], it also draws attention to the opportunities and difficulties businesses have when running their logistics departments in a fast-paced and cut-throat industry [24, 34]. It offers useful and doable advice on how to deal with the difficulties and seize the chances [8, 15]. Readers will obtain a thorough understanding of the connection between logistics operations and financial performance by reading this chapter [24]. Through effective and efficient logistics operations, they will also learn how to apply the knowledge and skills to their own businesses and enhance their financial performance [11, 34].

According to Ref. [37] a MIT1 study, businesses can increase cash processing speed, lower inventory levels, and increase return on capital. Establish a connection between supply chain procedures and operational and financial performance to improve cash cycles and customer satisfaction [34]. A survey of 1128 supply chain executives from 50 countries, representing a range of industries and company sizes, served as the basis for MIT1's study [35, 37]. Businesses that align their supply chain strategies with operational and financial performance indicators have a greater chance than their rivals of achieving improved return on capital, lower inventory levels, quicker cash-to-cash cycles, and higher customer satisfaction [35].

Strategic supplier partnerships, demand-driven operations, customer-centric innovation, integrated planning and execution, segment-specific service, and talent development are the other six critical supply chain practices that the study identified as driving operational and financial performance [35]. Businesses implement these supply chain strategies and match them with their objectives and those of their clients stand to gain operational and financial performance improvements [11, 25]. In order to evaluate supply chain performance and pinpoint areas for development, the study also offers businesses a framework and self-assessment tool. According to an Emerald study, businesses can increase their operational and financial performance by putting in place a logistics business continuity plan, which will also increase their competitiveness in the logistics market and increase their resilience to emergencies [20, 35].

A logistics business continuity plan is a plan that describes how the company will carry on offering distribution, warehousing, and transportation services in the case of crises or problems with the supply chain [12, 26]. Having a business continuity plan for logistics helps guarantee that, in an emergency, transportation operations continue to be safe and reliable [11, 12]. Logistics companies can use these processes, procedures, and tools to identify potential risks, adjust their operations accordingly, and ultimately minimize any negative effects on their business [15, 35]. By boosting their logistics competitiveness and strengthening their disaster immunity, businesses can enhance their operational and financial performance through the implementation of a logistics business continuity plan [28].

Logistics competitiveness refers to a company's capacity to surpass its rivals and deliver value to its clientele. This includes elements such as service excellence, dependability, adaptability, and creativity [37]. The capacity of a logistics company to endure and bounce back from unforeseen circumstances that could cause disruptions to its operations is known as disaster immunity. These consist of elements such as crisis management, emergency preparedness, risk assessment, and recovery tactics [15, 26]. In addition to improving financial performance indicators such as sales, profit, and return on investment, Emerald's study [11, 38] suggests that creating a logistics business continuity plan can enhance operational performance indicators such as order fulfillment, customer satisfaction, and on-time delivery [20]. Additionally, the study offers a framework and questionnaire to logistics firms so they can evaluate their current state of disaster immunity and logistics competitiveness and pinpoint areas in need of development.

Acknowledgements

As I wrote this book chapter, I wanted to take this opportunity to thank each and every one of you for your support. To begin with, I want to express my gratitude to my family for their unwavering love and support. Also, for making the required accommodations so that I could dedicate my entire time to this project. Without their understanding and assistance, this would not have been possible. Also deserving of my profound gratitude are my friends and colleagues, who throughout the writing process have provided invaluable counsel and insights. I am very much in gratitude for your support. For taking the time to read this chapter, I would like to thank the readers one last time. My goal is that the words on these pages will leave a lasting impression on you, encouraging you to make positive changes in your life. Thank you all so much for your assistance.

Conflict of interest

I assert that I have no conflicts of interest related to multiple interests held by a person or organization. These interests could potentially impact the purity of motives or decision-making process of one another. Financial, personal, and professional conflicts are among the many scenarios that can arise. As the writer, I want to emphasize that I do not possess any such potential conflicts. This assurance ensures that every one involved can make informed decisions and maintain integrity and trust. Hence, I affirm that there are no possible conflicts of interest associated with the information presented in this post or the knowledge it provides.

Other declarations

The only author of this chapter is Dr. Dennis Sebata, who carefully examined supply chain and logistics engineering from both a historical and a contemporary perspective. This chapter is a great resource for anyone looking to learn more about supply chain management and logistics because of the author's in-depth knowledge of the subject and background in the field, which have allowed him to offer thorough insights into the topic.

Author details

Dennis Sebata
Institute of Development Management (IDM) Botswana, Francistown, Botswana

*Address all correspondence to: densebj@gmail.com

References

[1] Murfield MLU, Boone CG, Rutner P, Thomas RW. Investigating logistics service quality in omni-channel retailing. International Journal of Physical Distribution & Logistics Management. 2017;**47**(4):263-296

[2] Ishfaq R, Defee CC, Gibson B, Raja U. Realignment of the physical distribution process in omni-channel fulfillment. International Journal of Physical Distribution & Logistics Management. 2016;**46**(6/7):543-561

[3] Melacini M, Perotti S, Rasini M, Tappia E. E-fulfilment and distribution in omni-channel retailing: A systematic literature review. International Journal of Physical Distribution & Logistics Management. 2018;**48**(4):391-414

[4] Branch BS. Research and Development activity and profitability: A distributed lag analysis. Journal of Political Economy [Internet]. 1974;**82**(5):999-1011. DOI: 10.1086/260252

[5] Pardalos PM, Wollenburg J, Holzapfel A. Retail logistics in the transition from multi-channel to omni-channel. International Journal of Physical Distribution & Logistics Management. 2016;**46**(6/7):562-583

[6] Frazelle EH. Supply Chain Strategy: The Logistics of Supply Chain Management [Internet]. New York, NY: McGraw-Hill eBooks; 2002. Available from: https://ci.nii.ac.jp/ncid/BA54446040

[7] Prajogo DI, Oke A, Olhager J. Supply chain processes. International Journal of Operations & Production Management [Internet]. 2016;**36**(2):220-238. DOI: 10.1108/ijopm-03-2014-0129

[8] Christopher M. Logistics and supply chain management: Strategies for reducing costs and improving services. Journal of the Operational Research Society [Internet]. 1994;**45**(11):1341. DOI: 10.2307/2583864

[9] Hofmann E, Solakivi T, Töyli J, Zinn M. Oil price shocks and the financial performance patterns of logistics service providers. Energy Economics [Internet]. 2018;**72**:290-306. DOI: 10.1016/j.eneco.2018.04.006

[10] Jaekel F. Fundamentals of logistics systems. In: Springer eBooks [Internet]. Berlin, NewYork: Springer; 2018. pp. 9-51. DOI: 10.1007/978-3-658-22837-8_2

[11] Issaoui Y, Khiat A, Bahnasse A, Ouajji H. Toward smart logistics: Engineering insights and emerging trends. Archives of Computational Methods in Engineering. 2020;**28**(4): 3183-3210. DOI: 10.1007/s11831-020-09494-2

[12] Rodrigues VS, Potter AT, Naim MM. The impact of logistics uncertainty on sustainable transport operations. International Journal of Physical Distribution & Logistics Management [Internet]. 2010;**40**(1/2):61-83. DOI: 10.1108/09600031011018046

[13] Naim MM, Aryee G, Potter AT. Determining a logistics provider's flexibility capability. International Journal of Production Economics [Internet]. 2010;**127**(1):39-45. DOI: 10.1016/j.ijpe.2010.04.011

[14] Martí L, Puertas R, Garcia LMT. The importance of the logistics performance index in international trade. Applied Economics

[Internet]. 2014;**46**(24):2982-2992. DOI: 10.1080/00036846.2014.916394

[15] Transport Engineering and Logistics [Internet]. TU Delft. Available from: https://www.tudelft.nl/3me/over/afdelingen/maritime-and-transport-technology/research/transport-engineering-and-logistics

[16] Winkelhaus S, Grosse EH. Logistics 4.0: A systematic review towards a new logistics system. International Journal of Production Research [Internet]. 2019;**58**(1):18-43. DOI: 10.1080/00207543.2019.1612964

[17] Woschank M, Pacher C. Program planning in the context of industrial logistics engineering education. Procedia Manufacturing [Internet]. 2020;**51**:1819-1824. DOI: 10.1016/j.promfg.2020.10.253

[18] Thomé AMT, Barbastefano RG, Scavarda LF, Reis JCGD, Amorim MPC. Industrial Engineering and Operations Management: XXVI IJCIEOM, Rio de Janeiro, Brazil, July 8-11, 2020. Sao Paulo: Springer Nature; 2020

[19] Zhong Y, Shi-Ming L, Bao M, Lv X. Dynamic network planning of underground logistics system on uncertainty graph. Mathematical Problems in Engineering [Internet]. 2019;**2019**:1-13. DOI: 10.1155/2019/1979275

[20] Datta PP, Roy R. Operations strategy for the effective delivery of integrated industrial product-service offerings. International Journal of Operations & Production Management [Internet]. 2011;**31**(5):579-603. DOI: 10.1108/01443571111126337

[21] Govindan K, Murugesan P, Zhu Q, Kannan D. Analysis of third party reverse logistics provider using interpretive structural modeling. International Journal of Production Economics [Internet]. 2012;**140**(1):204-211. DOI: 10.1016/j.ijpe.2012.01.043

[22] Rodrigues VS, Stantchev D, Potter AT, Naim MM, Whiteing A. Establishing a transport operation focused uncertainty model for the supply chain. International Journal of Physical Distribution & Logistics Management [Internet]. 2008;**38**(5):388-411. DOI: 10.1108/09600030810882807

[23] Demajorovic J, Augusto E, De Souza MTS. Reverse logistics of e-waste in developing countries: Challenges and prospects for the Brazilian model. Ambiente & Sociedade [Internet]. 2016;**19**(2):117-136. DOI: 10.1590/1809-4422asoc1415 45v1922016

[24] Katsaliaki K, Galetsi P, Kumar S. Supply chain disruptions and resilience: A major review and future research agenda. Annals of Operations Research [Internet]. 2021;**319**(1):965-1002. DOI: 10.1007/s10479-020-03912-1

[25] Ali A. In: Ali A, editor. Safety & Security in logistics. Medium. Pakistan. [Internet]; 2017. Available from: https://medium.com/@asad.behan/safety-security-in-logistics-c8bdfcfb8a76

[26] Yaacoub JPA, Noura H, Salman O, Chehab A. Robotics cyber security: Vulnerabilities, attacks, countermeasures, and recommendations. International Journal of Information Security [Internet]. 2021;**21**(1):115-158. DOI: 10.1007/s10207-021-00545-8

[27] Leung KH, Choy K, Siu PKY, Ho GTS, Lam HY, Lee CKM. A B2C e-commerce intelligent system for re-engineering the e-order fulfilment process. Expert Systems with Applications [Internet]. 2018;**91**:386-401. DOI: 10.1016/j.eswa.2017.09.026

[28] Lambert DM, Stock JR, Ellram LM. Fundamentals of Logistics [Internet]. Boston, MA: Irwin/McGraw-Hill; 1997. Available from: https://is.vstecb.cz/publication/26787?lang=en

[29] Nikolaou IE, Evangelinos K, Allan S. A reverse logistics social responsibility evaluation framework based on the triple bottom line approach. Journal of Cleaner Production [Internet]. 2013;**56**:173-184. DOI: 10.1016/j.jclepro.2011.12.009

[30] Liu CL, Lai PY. Impact of external integration capabilities of third-party logistics providers on their financial performance. The International Journal of Logistics Management [Internet]. 2016;**27**(2):263-283. DOI: 10.1108/ijlm-09-2014-0155

[31] Langford J. Logistics: Principles and Applications [Internet]. New York, NY: SOLE Press/McGraw-Hill; 1994. Available from: http://ci.nii.ac.jp/ncid/BA23797062

[32] Baah C, Amponsah KT, Issau K, Ofori D, Acquah ISK, Agyeman DO. Examining the interconnections between sustainable logistics practices, environmental reputation and financial performance: A mediation approach. Vision: The Journal of Business Perspective [Internet]. 2021;**25**(1):47-64. DOI: 10.1177/0972262920988805

[33] Govindan K, Soleimani H, Kannan D. Reverse logistics and closed-loop supply chain: A comprehensive review to explore the future. European Journal of Operational Research [Internet]. 2015;**240**(3):603-626. DOI: 10.1016/j.ejor.2014.07.012

[34] Buscher U, Neufeld JS, Lasch R, Schönberger J. Logistics Management: Contributions of the Section Logistics of the German Academic Association for Business Research. Dresden, Germany: Springer N; 2023

[35] Guidotti R, Monreale A, Ruggieri S, Turini F, Giannotti F, Pedreschi D. A survey of methods for explaining black box models. ACM Computing Surveys [Internet]. 2018;**51**(5):1-42. DOI: 10.1145/3236009

[36] Arumugam A, Teng TY, Hamid AB, Sukati I. A Logistics IT Strategy Firm Performance and Generation y Corporate Social Responsibility. Singapore: Partridge Publishing Singapore; 2020

[37] Shen T. Linking supply chain practices to operational and financial performance. In: Lapide L, Allen BS, editors. Cambridge, MA, USA: MIT Center for Transportation and Logistics [Internet]. 2020. Available from: https://ctl.mit.edu/sites/ctl.mit.edu/files/library/public/paper_SCMVersusPerformance_singh.pdf

[38] Yu P. 5 Steps to Master Financial Management in Logistics [Internet]. United States: IIENSTITU Bookboon; 2023. Available from: https://www.iienstitu.com/en/blog/5-steps-to-master-financial-management-in-logistics [Accessed: Jan 3, 2024]